HYDROMECHANICS

PART I

HYDROSTATICS

A TREATISE ON
HYDROMECHANICS

PART I
HYDROSTATICS

BY

W. H. BESANT, Sc.D., F.R.S.
LATE FELLOW AND LECTURER OF ST JOHN'S COLLEGE, CAMBRIDGE

AND

A. S. RAMSEY, M.A.
LATE FELLOW OF MAGDALENE COLLEGE, CAMBRIDGE

REVISED EDITION

CBS Publishers & Distributors Pvt. Ltd.

New Delhi • Bengaluru • Chennai • Kochi • Kolkata • Mumbai
Hyderabad • Nagpur • Patna • Pune • Vijayawada

ISBN: 81-239-1190-4

First Indian Edition: 1988
Reprint: 2002, 2004

This edition has been published in India by arrangement with
G. Bell & Sons Ltd., UK

Published by:
Satish Kumar Jain for CBS Publishers & Distributors Pvt. Ltd.,
4819/XI Prahlad Street, 24 Ansari Road, Daryaganj, New Delhi - 110002
delhi@cbspd.com, cbspubs@airtelmail.in • www.cbspd.com
Ph.: 23289259, 23266861, 23266867 • Fax: 011-23243014

Corporate Office: 204 FIE, Industrial Area, Patparganj, Delhi - 110 092
Ph: 49344934 • Fax: 011-49344935
E-mail: publishing@cbspd.com • publicity@cbspd.com

Branches:
- ***Bengaluru:*** 2975, 17th Cross, K.R. Road, Bansankari 2nd Stage, Bengaluru - 70 • Ph: +91-80-26771678/79 • Fax: +91-80-26771680 E-mail: cbsbng@gmail.com, bangalore@cbspd.com
- ***Chennai:*** No. 7, Subbaraya Street, Shenoy Nagar, Chennai - 600030 Ph: +91-44-26681266, 26680620 • Fax: +91-44-42032115 E-mail: chennai@cbspd.com
- ***Kochi:*** Ashana House, 39/1904, A.M. Thomas Road, Valanjambalam, Ernakulum, Kochi • Ph: +91-484-4059061-65 Fax: +91-484-4059065 • E-mail: cochin@cbspd.com
- ***Kolkata:*** 6-B, Ground Floor, Rameshwar Shaw Road, Kolkata - 700014 Ph: +91-33-22891126/7/8 • E-mail: kolkata@cbspd.com
- ***Mumbai:*** 83-C, Dr. E. Moses Road, Worli, Mumbai - 400018 Ph: +91-9833017933, 022-24902340/41 • E-mail: mumbai@cbspd.com

Representatives:
- Hyderabad: 0-9885175004
- Nagpur: 0-9021734563
- Patna: 0-9334159340
- Pune: 0-9623451994
- Vijayawada: 0-9000660880

Printed at:
Nazia Printers, Delhi

PREFACE TO THE NINTH EDITION

In preparing a new edition of this book for publication, attention has been given to the change in outlook in mathematical studies in Cambridge that began with the abolition of the order of merit in the Tripos. Hydrostatics is still a subject which all candidates are expected to study; but it belongs to the class of blind-alley subjects, and it is clearly not profitable for the average student to devote very much of his time either to the subject-matter or to working elaborate problems. In the interests of the average student, therefore, the amount of book-work has been substantially reduced, and a large number of examples have been removed from the book, while a few from recent Tripos papers have been inserted.

July 1925.

In this impression a few corrections have been made, particularly in Chapters VI and VII, but in other respects the book is unaltered. A. S. R.

May 1929.

CONTENTS

HYDROSTATICS

CHAPTER I

CHAPTER II

CHAPTER III

CHAPTER IV

CHAPTER V

CHAPTER VI

CHAPTER VII

CHAPTER VIII

HYDROSTATICS

CHAPTER I

1. We learn from common experience that such substances as air and water are characterised by the ease with which portions of their mass can be removed, and by their extreme divisibility. These properties are illustrated by various common facts ; if, for instance, we consider the ease with which fluids can be made to permeate each other, the extreme tenuity to which one fluid can be reduced by mixture with a large portion of another fluid, the rarefaction of air which can be effected by means of an air-pump, and other facts of a similar kind, it is clear that, practically, the divisibility of fluid is unlimited : we find, moreover, that in separating portions of fluids from each other, the resistance offered to the division is very slight, and in general almost inappreciable. By a generalisation from such observations, the conception naturally arises of a substance possessing in the highest degree these properties, which exist, in a greater or less degree, in every fluid with which we are acquainted, and hence we are led to the following

Definition of a Perfect Fluid

2. *A perfect fluid is an aggregation of particles which yield at once to the slightest effort made to separate them from each other.*

If then an indefinitely thin plane be made to divide such a fluid in any direction, no resistance will be offered to the division, and the pressure exerted by the fluid on the plane will be entirely normal to it ; that is, a perfect fluid is assumed to have no " viscosity," no property of the nature of friction.

The following fundamental property of a fluid is therefore obtained from the above definition.

The pressure of a perfect fluid is always normal to any surface with which it is in contact.

As a matter of fact, all fluids do more or less offer a resistance to separation or division, but, just as the idea of a rigid body is obtained from the observation of bodies in nature which only change form slightly on the application of great force, so is the idea of a perfect fluid obtained from our experiences of substances which possess the characteristics of extremely easy separability and apparently unlimited divisibility.

The following definition will include fluids of all degrees of viscosity.

A fluid is an aggregation of particles which yield to the slightest effort made to separate them from each other, if it be continued long enough.

Hence it follows that, in a viscous fluid at rest, there can be no tangential action, or shearing stress, and therefore, as in the case of a perfect fluid,

The pressure of a fluid at rest is always normal to any surface with which it is in contact.

Thus all propositions in Hydrostatics are true for all fluids whatever be the viscosity.

In Hydrodynamics it will be found that the equations of motion are considerably modified by taking account of the viscosity of a fluid.

3. Fluids are divided into Liquids and Gases ; the former, such as water and mercury, are not sensibly compressible except under very great pressures ; the latter are easily compressible, and expand freely if permitted to do so.

Hence the former are sometimes called inelastic, and the latter elastic fluids.

4. Fluids are acted upon by the force of gravity in the same way as solids ; with regard to liquids this is obvious ; and that air has weight can be shown directly by weighing a closed vessel, exhausted as far as possible : moreover, the phenomena of the tides show that fluids are subject to the attractive forces of the sun and moon as well as of the earth, and it is assumed, from these and other similar facts, that fluids of all kinds are subject to the law of gravitation, that is, that they attract, and are attracted by, all other portions of matter, in accordance with that law.

Measure of the Pressure of Fluids

5. Consider a mass of fluid at rest under the action of any forces, and let A be the area of a plane surface exposed to the action of the fluid, that is, in contact with it, and P the force which is required to counterbalance the action of the fluid upon A. If the action of the fluid upon A be uniform, then $\frac{P}{A}$ is the pressure on each unit of the area A. If the pressure be not uniform, it must be considered as varying continuously from point to point of the area A, and if ϖ be the force on a small portion a of the area about a given point, then $\frac{\varpi}{a}$ will approximately express the *rate* of pressure over a. When a is indefinitely diminished let $\frac{\varpi}{a}$ ultimately $=p$, then p is defined to be the measure of the pressure at the point considered, p being the force which would be exerted on a unit of area, if the rate of pressure over the unit were uniform and the same as at the point considered.

The force upon any small area a about a point, the pressure *at* which is p, is therefore $pa+\gamma$, where γ vanishes ultimately in comparison with pa when a (and consequently pa) vanishes.

6. *The pressure at any point of a fluid at rest is the same in every direction.*

This is the most important of the characteristic properties of a fluid ; it can be deduced from the fundamental property of a fluid in the following manner.

If we consider the equilibrium of a small tetrahedron of fluid, we observe that the pressures on its faces, and the impressed force on its mass, form a system of equilibrating forces.

The former forces depending on the areas of the faces vary as the square, and the latter depending on the volume and density varies as the cube of one of the edges of the solid, which is considered to be homogeneous, and therefore supposing the solid indefinitely diminished, while it retains always a similar form, the latter force vanishes in comparison with the pressures on the faces ; and these pressures consequently form of themselves a system of forces in equilibrium.

Let p, p' be the rates of the pressure on the faces ABC, BCD, and resolve the forces parallel to the edge AD; then,

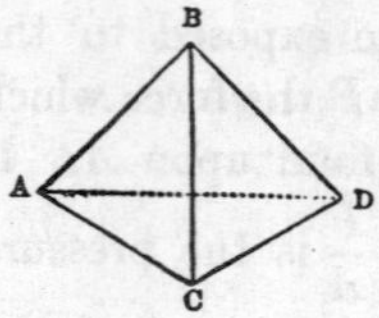

since the projections of the areas ABC, BCD on a plane perpendicular to AD are the same (each equal to a suppose) we have ultimately,

$$pa = p'a,$$

or $$p = p'.$$

And similarly it may be shown that the pressures on the other two faces are each equal to p or p'.

As the tetrahedron may be taken with its faces in any direction, it follows that the pressure at a point is the same in every direction.

This proposition is also true if the fluid be in motion, for by D'Alembert's Principle the reversed effective forces and the impressed forces which act upon the mass of fluid must balance the pressures on its faces, and the effective forces are of the same order of small quantities as the impressed forces and vanish in comparison with the pressures.

Transmission of Fluid Pressure

7. *Any pressure, or additional pressure, applied to the surface or to any other part of a liquid at rest, is transmitted equally to all parts of the liquid.*

This property of liquids is a direct result of experiment, and, as such, is sometimes assumed. It is, however, deducible from the definition of a fluid.

Let P be a point in the surface of a liquid at rest, and Q any other point in the liquid; about the straight line PQ describe a cylinder, of very small radius, bounded by the surface at P and by a plane through Q, perpendicular to QP.

If the pressure at P be increased by p, the additional force on the cylinder, resolved in the direction of its axis, is pa, a being the area of the section of the cylinder perpendicular to its axis, and this must be counteracted by an equal force pa at Q in the direction QP, since the pressure of the liquid on the curved surface is perpendicular to the axis. The pressure at Q is therefore increased by p.

If the straight line PQ do not lie entirely in the liquid, P and Q can be connected by a number of straight lines, all lying in the liquid, and a repetition of the above reasoning will show that the pressure p is transmitted, unchanged, to the point Q.

8. In consequence of this property, a mass of liquid can be used as a " machine " for the purpose of multiplying power.

Thus, if in a closed vessel full of water two apertures be made and pistons A, A' fitted in them, any force P applied to one piston must be counteracted by a force P' on the other piston, such that $P' : P$ in the ratio of the area $A' : A$, for the increased rate of pressure at every point of A is transmitted to every point of A', and the force upon A' depends therefore upon its area.*

The action between the two is analogous to the action of a lever, and it is clear that by increasing A' and diminishing A, we can make the ratio $P' : P$ as large as we please.

9. The pressure of a gaseous fluid is found to depend upon its density and temperature, as well as upon the nature of the fluid itself.

When the temperature is constant, experiment shows that the pressure varies inversely as the space occupied by the fluid, that is, directly as its density.

This law was first stated by Boyle, but it is a consequence of the more general law that the pressure of a mixture of gases that do not act chemically on each other is the sum of the pressures the gases would exert if they filled the containing vessel separately. For doubling the quantity of gas in the vessel would double the pressure, and a similar proportionate change of pressure would take place for any other change of quantity.

Hence if ρ be the density of a certain quantity of a gaseous fluid, and p its pressure, then, as long as the temperature remains the same,

$$p=k\rho,$$

where k is a constant, to be determined experimentally for the fluid at a given temperature.

If v be the volume of the gas at the pressure p, and v' the volume at the pressure p',

$$pv=p'v',$$

or pv is constant for a given temperature.

10. The **Elasticity** of a fluid is measured by the ratio of a

* Bramah's press is an instance of the practical use of this property of liquids.

small increase of pressure to the cubical compression produced by it.

If v be the volume, the small cubical compression is $-\frac{dv}{v}$, and the measure of the elasticity is

$$-v\frac{dp}{dv}.$$

In the case of a gas at constant temperature pv is constant, and

$$\therefore p+v\frac{dp}{dv}=0,$$

so that the measure of the elasticity is equal to that of the pressure.

If the relation between the elasticity and the pressure is given, we can deduce the relation between the pressure and the volume.

For instance, if we can imagine the existence of a fluid in which the elasticity is double the pressure, we have

$$-v\frac{dp}{dv}=2p,$$

from which it follows that pv^2 is constant.

Measures of Weight, Mass, and Density

11. The weight, mass, and density of a fluid are measured in the same way as for solid bodies.

If W be the weight of a mass M of fluid, then, in accordance with the usual conventions which define the units of mass and force,

$$W=Mg.$$

If V be the volume of the mass M of fluid of density ρ, then

$$M=\rho V,$$

and

$$\therefore W=g\rho V.$$

For the standard substance, $\rho=1$, and therefore the unit of mass is the mass of the unit of volume of the standard substance.

If the unit of mass is a pound, the equation, $W=Mg$, shows that the action of gravity on a pound is equivalent to g units of force. The unit of force is therefore, roughly, equal to the weight of half an ounce, and it is called the Poundal.

12. In the previous articles no account has been taken of fluids in which the density is variable; but it is easy to conceive the density of a mass of liquid varying continuously from point to point, and it will be hereafter found that a mass of elastic fluid, at rest under the action of gravity, and having a constant temperature throughout, is necessarily heterogeneous: the density at a point of a fluid must therefore be measured in the same way as the pressure at a point, or any other continuously varying quantity.

Measure of the density at any point of a heterogeneous fluid.

Let m be the mass of a volume v of fluid enclosing a given point, and suppose ρ the density of a homogeneous fluid such that the mass of a volume v is equal to m, or such that

$$m = \rho v;$$

then ρ may be defined as the mean density of the portion v of the heterogeneous fluid, and the ultimate value of ρ when v is indefinitely diminished, supposing it always to enclose the point, is the density of the fluid at that point.

EXAMPLES

(In these Examples g is taken to be 32, when a foot and a second are units.)

1. $ABCD$ is a rectangular area subject to fluid pressure; AB is a fixed line, and the pressure on the area is a given function (P) of the length BC (x); prove that the pressure at any point of CD is $\dfrac{dP}{a\,dx}$, where $a = AB$.

If A be a fixed point, and AB, AD fixed in direction, and if $AB = x$ and $AD = y$, the pressure at $C = \dfrac{\partial^2 P}{\partial x \partial y}$.

2. In the equation $W = g\rho V$, if the unit of force be 100 lb. weight, the unit of length 2 feet, and the unit of time $\frac{1}{4}$th of a second, find the density of water.

3. If a minute be the unit of time, and a yard the unit of space, and if 15 cubic inches of the standard substance contain 25 oz., determine the unit of force.

4. In the equation, $W = g\rho V$, the number of seconds in the unit of time is equal to the number of feet in the unit of length, the unit of force is 750 lb. weight, and a cubic foot of the standard substance contains 13500 ounces; find the unit of time.

5. A velocity of 4 feet per second is the unit of velocity; water is the standard substance and the unit of force is 125 lb. weight; find the units of time and length.

6. The number expressing the weight of a cubic foot of water is $\frac{1}{10}$th of that expressing its volume, $\frac{1}{8}$th of that expressing its mass, and $\frac{1}{100}$th of the number expressing the work done in lifting it 1 foot. Find the units of length, mass, and time.

7. If a feet and b seconds be the units of space and time, and the density of water the standard density, find the relation between a and b in order that the equation, $W = g\rho V$, may give the weight of a substance in pounds.

8. A velocity of 8 feet per second is the unit of velocity, the unit of accéleration is that of a falling body, and the unit of mass is a ton; find the density of water.

9. The density at any point of a liquid, contained in a cone having its axis vertical and vertex downwards, is greater than the density at the surface by a quantity varying as the depth of the point. Show that the density of the liquid when mixed up so as to be uniform will be that of the liquid originally at the depth of one-fourth of the axis of the cone.

10. From a vessel full of liquid of density ϱ is removed $1/n$th of the contents, and it is filled up with liquid of density σ. If this operation be repeated m times, find the resulting density in the vessel.

Deduce the density in a vessel of volume V, originally filled with liquid of density ϱ, after a volume U of liquid of density σ has dripped into it by infinitesimal drops.

CHAPTER II

THE CONDITIONS OF THE EQUILIBRIUM OF FLUIDS

13. Taking the most general case, suppose a mass of fluid, elastic or non-elastic, homogeneous or heterogeneous, to be at rest under the action of given forces, and let it be required to determine the conditions of equilibrium, and the pressure at any point.

Let x, y, z be the co-ordinates referred to rectangular axes, of any point P in the fluid, and let Q be a point near it, so taken that PQ is parallel to the axis of x.

Take $x+\delta x$, y, z as the co-ordinates of Q; about PQ describe a small prism or cylinder terminated by planes perpendicular to PQ.

Let α be the area of the section of the cylinder perpendicular to its axis, p the pressure at P, and $p+\delta p$ the pressure at Q.

Then α may be taken so small that the thrust on the plane end at P is approximately $p\alpha$, the difference being of a higher order of smallness.

Similarly the thrust on the plane end at Q may be taken to be

$$(p+\delta p)\alpha.$$

If ρ be the mean density of the cylinder PQ, its mass $=\rho\alpha\delta x$, and $X\rho\alpha\delta x$ will represent the force on PQ parallel to its axis, if $X\delta m$, $Y\delta m$, $Z\delta m$ be the components of the forces acting on a particle δm of fluid at the point (x, y, z).

Hence, for the equilibrium of PQ,

$$(p+\delta p)\alpha-p\alpha=X\rho\alpha\delta x,$$

or

$$\delta p=\rho X\delta x.$$

Proceeding to the limit when δx, and therefore δp, is indefinitely diminished, ρ will be the density at P, and we obtain

$$\frac{\partial p}{\partial x}=\rho X.$$

By a similar process,

$$\frac{\partial p}{\partial y}=\rho Y,$$

$$\frac{\partial p}{\partial z}=\rho Z.$$

But
$$dp=\frac{\partial p}{\partial x}dx+\frac{\partial p}{\partial y}dy+\frac{\partial p}{\partial z}dz\,;$$

$$\therefore\quad dp=\rho(Xdx+Ydy+Zdz)\quad . \quad . \quad . \quad (1)$$

the equation which determines the pressure.

14. We now consider what condition must be satisfied by a given distribution of force in order that it may be capable of maintaining a fluid in equilibrium. The pressure is clearly a function of the independent variables x, y, *and* z, and we know that

$$\frac{\partial^2 p}{\partial y\partial z}=\frac{\partial^2 p}{\partial z\partial y},\quad \frac{\partial^2 p}{\partial z\partial x}=\frac{\partial^2 p}{\partial x\partial z},\quad \frac{\partial^2 p}{\partial x\partial y}=\frac{\partial^2 p}{\partial y\partial x}.$$

Hence we obtain from the preceding equations,

$$\left.\begin{aligned}\frac{\partial}{\partial y}(\rho Z)&=\frac{\partial}{\partial z}(\rho Y)\\ \frac{\partial}{\partial z}(\rho X)&=\frac{\partial}{\partial x}(\rho Z)\\ \frac{\partial}{\partial x}(\rho Y)&=\frac{\partial}{\partial y}(\rho X)\end{aligned}\right\}\quad . \quad . \quad . \quad (2)$$

Performing the operations indicated we have

$$Z\frac{\partial\rho}{\partial y}-Y\frac{\partial\rho}{\partial z}=\rho\left(\frac{\partial Y}{\partial z}-\frac{\partial Z}{\partial y}\right),$$

$$X\frac{\partial\rho}{\partial z}-Z\frac{\partial\rho}{\partial x}=\rho\left(\frac{\partial Z}{\partial x}-\frac{\partial X}{\partial z}\right),$$

$$Y\frac{\partial\rho}{\partial x}-X\frac{\partial\rho}{\partial y}=\rho\left(\frac{\partial X}{\partial y}-\frac{\partial Y}{\partial x}\right).$$

Multiplying by X, Y, Z, and adding, we obtain

$$X\left(\frac{\partial Y}{\partial z}-\frac{\partial Z}{\partial y}\right)+Y\left(\frac{\partial Z}{\partial x}-\frac{\partial X}{\partial z}\right)+Z\left(\frac{\partial X}{\partial y}-\frac{\partial Y}{\partial x}\right)=0.\quad . \quad . \quad (3)$$

as a necessary condition of equilibrium.

The geometrical interpretation of this equation is that the lines of force,

$$\frac{dx}{X}=\frac{dy}{Y}=\frac{dz}{Z},$$

can be intersected orthogonally by a system of surfaces.

15. Homogeneous Liquids. If the fluid be homogeneous and incompressible, ρ is constant, and it follows from (1) that $Xdx+Ydy+Zdz$ must be a perfect differential in order that equilibrium may be possible.

In other words, the system of forces must be a conservative system, and the forces can be represented by the space-variations of a potential function.

We then have, if V be the potential function,

$$dp=-\rho dV,$$

and $$\therefore\ \frac{p}{\rho}+V=C.$$

If, for instance, the forces tend to or from fixed centres and are functions of the distances from those centres, we have

$$X=\Sigma\left\{\phi(r)\frac{x-a}{r}\right\},\quad Y=\Sigma\left\{\phi(r)\frac{y-b}{r}\right\},\quad Z=\Sigma\left\{\phi(r)\frac{z-c}{r}\right\},$$

where (a, b, c) are co-ordinates of the centre to which the force $\phi(r)$ tends.

Now $$r^2=(x-a)^2+(y-b)^2+(z-c)^2,$$

$$\therefore\ Xdx+Ydy+Zdz=\Sigma\phi(r)dr,$$

and $$dp=\rho\Sigma\phi(r)dr.$$

In this case, since

$$\frac{\partial X}{\partial y}=\Sigma\left\{\phi'(r)\frac{x-a}{r}\frac{y-b}{r}-\phi(r)\frac{x-a}{r^2}\frac{y-b}{r}\right\},$$

and $$\frac{\partial Y}{\partial x}=\Sigma\left\{\phi'(r)\frac{y-b}{r}\frac{x-a}{r}-\phi(r)\frac{y-b}{r^2}\frac{x-a}{r}\right\},$$

it is obvious that the equation (3) is always satisfied, but it is not to be inferred that the equilibrium of a heterogeneous fluid is always possible with such a system of forces.

When the density is constant, the equations (2) become

$$\frac{\partial Z}{\partial y}=\frac{\partial Y}{\partial z},\quad \frac{\partial X}{\partial z}=\frac{\partial Z}{\partial x},\quad \frac{\partial X}{\partial y}=\frac{\partial Y}{\partial x},$$

which are in this case always satisfied, and therefore the equilibrium of a homogeneous fluid under the action of such forces is always possible.

16. Heterogeneous Fluids. If the law of density be prescribed, that is, if ρ be a given function of x, y, z, the conditions to be satisfied in order that a given distribution of force, represented by X, Y, Z, may maintain the fluid in equilibrium are the equations (2).

17. Elastic Fluids. When the fluid is elastic, an additional condition is introduced, for, as we have seen in Chapter I, if the temperature be constant,

$$p=k\rho;$$

$$\therefore \frac{dp}{p}=\frac{1}{k}(Xdx+Ydy+Zdz) \quad . \quad . \quad . \quad (4)$$

If the forces are derivable from a potential V, *i.e.* if

$$Xdx+Ydy+Zdz$$

be a perfect differential $-dV$,

$$k\frac{dp}{p}=-dV,$$

$$\therefore k\log\frac{p}{C}=-V,$$

and

$$\text{or} \quad p=Ce^{-\frac{V}{k}}, \quad \text{and} \quad \rho=\frac{C}{k}e^{-\frac{V}{k}}.$$

When the forces tend to fixed centres and are functions of the distances, Art. 15, this equation takes the form

$$k\frac{dp}{p}=\Sigma\phi(r)dr,$$

and p can be determined.

If the temperature be variable, the relation between the pressure, density, and temperature is found to be

$$p=k\rho(1+\alpha t),$$

where t is the temperature, measured by a Centigrade thermometer, and $\alpha=\cdot003665$.

From this we obtain

$$p=k\rho\alpha\left\{\frac{1}{\alpha}+t\right\}=K\rho T,$$

where $$K=k\alpha, \quad \text{and} \quad T=\frac{1}{\alpha}+t.$$

T is called the absolute temperature, the zero of which is $-273°$ C.

In this case $$\frac{dp}{p}=\frac{Xdx+Ydy+Zdz}{KT},$$

and therefore T must be a function of x, y, z.

In any of these cases, if the pressure at any particular point be given, the constant can be determined.

In the case of elastic fluids, if the mass of fluid and the space within which it is contained be given, the constant is determined.

18. The equation for determining p may also be obtained in the following manner.

Let PQ be the axis of a very small cylinder bounded by planes perpendicular to PQ.

Let p and $p+\delta p$ be the pressures at P and Q, a the areal section, and δs the length of PQ. Then, if $S\delta m$ be the component, in the direction PQ, of the forces acting on an element δm,

$$(p+\delta p)a-pa=\varrho aS\delta s,$$

and therefore, proceeding to the limit,

$$dp=\varrho Sds.$$

That is, the rate of increase of the pressure in any direction is equal to the product of the density and the resolved part of the force in that direction.

If x, y, z be the co-ordinates of P, and X, Y, Z the components of S parallel to the axes,

$$S=X\frac{dx}{ds}+Y\frac{dy}{ds}+Z\frac{dz}{ds},$$

and $\therefore\ dp=\varrho(Xdx+Ydy+Zdz)$ as in Art. 13.

If the position of P be given by the cylindrical co-ordinates r, θ, and z, and if P, T, Z be the components of S in the directions of r, θ, z,

$$S=P\frac{dr}{ds}+T\frac{rd\theta}{ds}+Z\frac{dz}{ds},$$

and the equation for p becomes

$$dp=\varrho\{Pdr+Trd\theta+Zdz\}.$$

Again, if the position of P be given by the ordinary polar co-ordinates r, θ, ϕ, and if the components of the force be R, N, and T, in the directions of r, of the perpendicular to the plane of the angle θ, and of the line perpendicular to r in that plane, it will be found that

$$\frac{dp}{\varrho}=Rdr+Nr\sin\theta d\phi+Trd\theta.$$

In a similar manner the expression for dp may be obtained for any other system of co-ordinates.

19. Surfaces of Equal Pressure. In all cases, in which the equilibrium of the fluid is possible, we obtain by integration

$$p=\phi(x, y, z).$$

If p be constant, $\qquad \phi(x, y, z)=p \qquad . \quad . \quad . \quad . \qquad (5)$

is the equation to a surface at all points of which the pressure is constant, and by giving different values to p we obtain a series of surfaces of equal pressure, and the external surface, or free surface, is obtained by making p equal to the pressure external to the fluid.

If the external pressure be zero, the free surface is therefore

$$\phi(x, y, z)=0.$$

The quantities $\frac{\partial\phi}{\partial x}, \frac{\partial\phi}{\partial y}, \frac{\partial\phi}{\partial z},$

which are proportional to the direction-cosines of the normal at the point (x, y, z) of the surface (5), are equal to

$$\frac{\partial p}{\partial x}, \frac{\partial p}{\partial y}, \frac{\partial p}{\partial z},$$

respectively, *i.e.* to $\rho X, \rho Y, \rho Z$, and are therefore proportional to X, Y, Z.

Hence the resultant force at any point is in direction of the normal to the surface of equal pressure passing through the point.

The surfaces of equal pressure are therefore the surfaces intersecting orthogonally the lines of force.

It follows from this result that a necessary condition of equilibrium is the existence of a system of surfaces orthogonal to the lines of force, a conclusion derivable also from the equation (3) of Art. 14, for that equation is the known analytical condition requisite for the existence of such a system.

20. If the fluid be a homogeneous liquid, that is, if ρ is constant, $Xdx+Ydy+Zdz$ must be a perfect differential, or in other words, the system of forces must be a conservative system.

In general, when the force-system is conservative, ρ must be a function of the potential V.

For $dp=-\rho dV$, and, dp being a perfect differential, ρ must be a function of V; hence V, and therefore ρ, is a function of p, and surfaces of equal pressure are equipotential surfaces, and are also surfaces of equal density.*

* These results may also be obtained in the following manner:

Consider two consecutive surfaces of equal pressure, containing between them a stratum of fluid, and let a small circle be described about a point P in one surface, and a portion of the fluid cut out by normals through the circumference. The portion of fluid is kept at rest by the impressed force, and by the pressures on its ends and on its circumference. Being very nearly a small cylinder, and the pressures at all points of its circumference being equal, the difference of the pressures on its two faces must be due to the force, which must therefore act in the same direction as these pressures, *i.e.* in direction of the normal at P.

If the forces are derivable from a potential, the resulting force is perpendicular

If the fluid be elastic and the temperature variable

$$\frac{dp}{p}=-\frac{dV}{KT}.$$

Hence by a similar process of reasoning T is a function of p, and surfaces of equal pressure are also surfaces of equal temperature.

If however $Xdx+Ydy+Zdz$ be not a perfect differential, these surfaces will not in general coincide.

Let the fluid be heterogeneous and incompressible; then the surfaces of equal pressure and of equal density are given respectively by the equations

$$dp=0, \quad d\rho=0,$$

or

$$\left.\begin{aligned} Xdx+Ydy+Zdz&=0\\ \frac{\partial\rho}{\partial x}dx+\frac{\partial\rho}{\partial y}dy+\frac{\partial\rho}{\partial z}dz&=0\end{aligned}\right\} \quad . \quad . \quad . \quad (6)$$

These then are the differential equations of surfaces which by their intersections determine curves of equal pressure and density.

From (6) we obtain

$$\frac{dx}{Z\frac{\partial\rho}{\partial y}-Y\frac{\partial\rho}{\partial z}}=\frac{dy}{X\frac{\partial\rho}{\partial z}-Z\frac{\partial\rho}{\partial x}}=\frac{dz}{Y\frac{\partial\rho}{\partial x}-X\frac{\partial\rho}{\partial y}} \quad . \quad . \quad (7)$$

But from the conditions of equilibrium we have

$$\rho\frac{\partial X}{\partial y}+X\frac{\partial\rho}{\partial y}=\rho\frac{\partial Y}{\partial x}+Y\frac{\partial\rho}{\partial x},$$

$$\rho\frac{\partial Y}{\partial z}+Y\frac{\partial\rho}{\partial z}=\rho\frac{\partial Z}{\partial y}+Z\frac{\partial\rho}{\partial y},$$

$$\rho\frac{\partial Z}{\partial x}+Z\frac{\partial\rho}{\partial x}=\rho\frac{\partial X}{\partial z}+X\frac{\partial\rho}{\partial z},$$

and therefore the equations (7) become

$$\frac{dx}{\frac{\partial Z}{\partial y}-\frac{\partial Y}{\partial z}}=\frac{dy}{\frac{\partial X}{\partial z}-\frac{\partial Z}{\partial x}}=\frac{dz}{\frac{\partial Y}{\partial x}-\frac{\partial X}{\partial y}} \quad . \quad . \quad . \quad (8)$$

the differential equations of the curves of equal pressure and density

to the equipotential surfaces, and the surfaces of equal pressure are therefore identical with the equipotential surfaces.

Again, considering the equilibrium of the elemental cylinder, the force acting upon it, per unit of mass, is equal to the difference of potentials divided by the distance between the surfaces of equal pressure, and as the mass of the element is directly proportional to this distance, it follows that the density must be constant, that is, the surfaces of equal pressure are also surfaces of equal density.

21. We shall now show how to obtain the fundamental pressure equation by considering the equilibrium of a finite mass of fluid.

Let S be any closed surface drawn in the fluid, and l, m, n the direction-cosines of the normal at any point drawn outwards. The conditions of equilibrium of the mass of fluid within the surface S are summarised in the statement that the normal pressures on the boundary must counterbalance the effect of the given forces acting throughout the mass. Thus by resolving parallel to the axes we get three equations of the type

$$\iint lpdS = \iiint \rho X dxdydz \quad . \quad . \quad . \quad (9)$$

and by taking moments about the axes we get three other equations of the type

$$\iint p(ny-mz)dS = \iiint \rho(yZ-zY)dxdydz \quad . \quad . \quad (10)$$

where the double integrations extend to the whole surface S and the triple integrations are throughout the space enclosed.

Now by Green's Theorem,* we have

$$\iint lpdS = \iiint \frac{\partial p}{\partial x} dxdydz$$

and
$$\iint p(ny-mz)dS = \iiint \left(y\frac{\partial p}{\partial z} - z\frac{\partial p}{\partial y}\right)dxdydz,$$

so that (9) and (10) become

$$\iiint \left(\frac{\partial p}{\partial x} - \rho X\right)dxdydz = 0$$

and
$$\iiint \left\{y\left(\frac{\partial p}{\partial z} - \rho Z\right) - z\left(\frac{\partial p}{\partial y} - \rho Y\right)\right\}dxdydz = 0\,;$$

and there are two other pairs of like equations.

Since these equations must be satisfied for all ranges of integration in the fluid, it is clear that the necessary and sufficient conditions of equilibrium are

$$\frac{\partial p}{\partial x} = \rho X, \quad \frac{\partial p}{\partial y} = \rho Y, \quad \frac{\partial p}{\partial z} = \rho Z.$$

It is to be noted that since a perfect fluid is incapable of resisting shearing stress there can be no such stresses in a mass of fluid in equilibrium, and therefore it follows that the equations obtained by taking moments about the axes will of necessity be satisfied whenever the equations obtained by resolving parallel to the axes

* *Vide* any *Cours d'Analyse, e.g.* de la Vallée Poussin, t. i. p. 381 (4th ed.).

are satisfied. For in equilibrium the latter equations are true for any portion of the fluid finite or infinitesimal, and this balancing of forces ensures that the equations of moments are true also.

22. We can also prove that $\rho(Xdx+Ydy+Zdz)$ must be a perfect differential, by condensing the equilibrium of a spherical element of fluid.

For the pressures of the fluid on the surface of the element are all in direction of its centre, and therefore the moment of the acting forces about the centre must vanish.

Let x, y, z be co-ordinates of the centre, and $x+\alpha$, $y+\beta$, $z+\gamma$ of any point inside the small sphere.

Then, ρ being the density at the centre, the expression $\Sigma dm(Z\beta-Y\gamma)$ becomes

$$\iiint d\alpha d\beta d\gamma\left(\rho+\frac{\partial\rho}{\partial x}\alpha+\frac{\partial\rho}{\partial y}\beta+\frac{\partial\rho}{\partial z}\gamma\right)\left\{\beta\left(Z+\frac{\partial Z}{\partial x}\alpha+\frac{\partial Z}{\partial y}\beta+\frac{\partial Z}{\partial z}\gamma\right)\right.$$
$$\left.-\gamma\left(Y+\frac{\partial Y}{\partial x}\alpha+\frac{\partial Y}{\partial y}\beta+\frac{\partial Y}{\partial z}\gamma\right)\right\}.$$

Now $\iiint \alpha d\alpha d\beta d\gamma=0$, the centre of the sphere being the centre of gravity of the volume, $\iiint\beta\gamma d\alpha d\beta d\gamma=0$, etc., and, if $d\tau=d\alpha d\beta d\gamma$,

$$\iiint\alpha^2d\tau=\iiint\beta^2d\tau=\iiint\gamma^2d\tau=\tfrac{1}{3}\iiint(\alpha^2+\beta^2+\gamma^2)d\tau$$
$$=\tfrac{1}{3}\cdot\int_0^r 4\pi r'^4dr'=\tfrac{4}{15}\pi r^5.$$

The expression for the moment then becomes, neglecting higher powers of α, β, γ,

$$\left\{\frac{\partial}{\partial y}(\rho Z)-\frac{\partial}{\partial z}(\rho Y)\right\}\frac{4\pi r^5}{15},$$

and, in order that this may be evanescent, we must have

$$\frac{\partial}{\partial y}(\rho Z)=\frac{\partial}{\partial z}(\rho Y).$$

23. Fluid at rest under the action of gravity.

Taking the axis of z vertical, and measuring z downwards,

$$X=0,\quad Y=0,\quad Z=g,$$

and the equation (1) of Art. 13 becomes

$$dp=g\rho dz,$$

an equation which may also be obtained directly by considering the equilibrium of a small vertical cylinder.

In the case of homogeneous liquid,

$$p=g\rho z+C,$$

and the surfaces of equal pressure are horizontal planes.

Hence the free surface is a horizontal plane, and, taking the origin in the free surface, and Π as the external pressure,

$$p=g\rho z+\Pi.$$

If there be no pressure on the free surface,

$$p=g\rho z,$$

or *the pressure at any point is proportional to the depth below the surface.*

In the case of heterogeneous liquid, the equation

$$dp=g\rho dz$$

shows that ρ must be a function of z. The density and pressure are therefore constant for all points in the same horizontal plane.

As an example, let $\rho \propto z^n=\mu z^n$,

then $$p=g\mu\frac{z^{n+1}}{n+1}+\Pi.$$

24. Elastic fluid at rest under the action of gravity.

In this case, $$p=k\rho,$$

and $$\frac{dp}{p}=\frac{g}{k}dz,$$

$$\therefore \log\frac{p}{C}=\frac{gz}{k} \text{ and } p=Ce^{\frac{gz}{k}}.$$

The surfaces of equal pressure are in this case also horizontal planes, and the constant C must be determined by a knowledge of the pressure for a given value of z, or by some other fact in connection with the particular case.

EXAMPLE. *A closed cylinder, the axis of which is vertical, contains a given mass of air.*

Measuring z from the top of the cylinder,

$$\varrho=\frac{p}{k}=\frac{C}{k}e^{\frac{gz}{k}};$$

$\therefore$ if M be the given mass, a the radius, and h the height of the cylinder,

$$M=\int_0^h \varrho\pi a^2 dz=\pi a^2\frac{C}{g}(e^{\frac{gh}{k}}-1),$$

whence C is determined.

25. *Illustrations of the use of the general equation.*

(1) Let a given volume V of liquid be acted upon by forces

$$-\frac{\mu x}{a^2},\ -\frac{\mu y}{b^2},\ -\frac{\mu z}{c^2},$$

respectively parallel to the axes;

then $$dp=\varrho\left(-\frac{\mu x}{a^2}dx-\frac{\mu y}{b^2}dy-\frac{\mu z}{c^2}dz\right),$$

and $$p=C-\frac{\mu\varrho}{2}\left(\frac{x^2}{a^2}+\frac{y^2}{b^2}+\frac{z^2}{c^2}\right).$$

The surfaces of equal pressure are therefore similar ellipsoids, and the equation to the free surface is

$$\frac{x^2}{a^2}+\frac{y^2}{b^2}+\frac{z^2}{c^2}=\frac{2C}{\mu\varrho},$$

assuming that there is no external pressure.

The condition which determines the constant is that the volume of the fluid is given, and we have

$$V=\tfrac{4}{3}\pi abc.\left(\frac{2C}{\mu\varrho}\right)^{\frac{3}{2}},$$

and $$C=\frac{\mu\varrho}{2}.\left(\frac{3V}{4\pi abc}\right)^{\frac{2}{3}}.$$

(2) *A given volume of liquid is at rest on a fixed plane, under the action of a force, to a fixed point in the plane, varying as the distance.*

Taking the fixed point as origin, the expression for the pressure at any point is

$$p=C-\tfrac{1}{2}\mu\varrho(x^2+y^2+z^2)=C-\tfrac{1}{2}\mu\varrho r^2,$$

where r is the distance from the origin; and if $\frac{2}{3}\pi a^3$ be the given volume, the free surface is a hemisphere of radius a, and

$$p=\tfrac{1}{2}\mu\varrho(a^2-r^2).$$

The portion of the plane in contact with fluid is a circle of radius a, and therefore the pressure upon it

$$=\int_0^{2\pi}\int_0^a pr\ d\theta\ dr$$

$$=\tfrac{1}{4}\pi\mu\varrho a^4.$$

This result may be written in the form $\mu\frac{3}{8}a\cdot\frac{2}{3}\pi\varrho a^3$, which is the expression for the attraction on the whole mass of fluid, supposed to be condensed into a material particle at its centre of gravity, and might in fact have been at once obtained by considering that the fluid is kept at rest by the attraction to the centre of force and the reaction of the plane.

(3) *A given volume of heavy liquid is at rest under the action of a force to a fixed point varying as the distance from that point.*

Take the fixed point as origin, and measure z vertically downwards;

then $$X=-\mu x, Y=-\mu y, \text{ and } Z=g-\mu z;$$

$$\therefore\ dp=\varrho\{-\mu xdx-\mu ydy+(g-\mu z)dz\},$$

and $$\frac{p}{\varrho}=C-\mu\frac{x^2+y^2+z^2}{2}+gz.$$

The surfaces of equal pressure are spheres, and the free surface, supposing the external pressure zero, is given by the equation

$$x^2+y^2+z^2-\frac{2g}{\mu}z=\frac{2C}{\mu}.$$

The volume of this sphere is

$$\tfrac{4}{3}\pi\left(\frac{2C}{\mu}+\frac{g^2}{\mu^2}\right)^{\frac{3}{2}};$$

equating this to the given volume, the constant C is determined, and the pressure at any point is then given in terms of r and z.

Rotating Fluids.

26. If a quantity of fluid revolve uniformly and without any relative displacement of its particles (*i.e.* as if rigid) about a fixed axis, the preceding equations will enable us to determine the pressure at any point, and the nature of the surfaces of equal pressure.

For, in such cases of relative equilibrium, every particle of the fluid moves uniformly in a circle, and the resultant of the external forces acting on any particle m of the fluid, and of the fluid pressure upon it, must be equal to a force $m\omega^2 r$ *towards* the axis, ω being the angular velocity, and r the distance of m from the axis; it follows therefore that the external forces, combined with the fluid pressures and forces $m\omega^2 r$ acting *from* the axis, form a system in statical equilibrium, to which the equations of the previous articles are applicable.

A mass of homogeneous liquid, contained in a vessel, revolves uniformly about a vertical axis; required to determine the pressure at any point, and the surfaces of equal pressure.

Take the vertical axis as the axis of z; then, resolving the force $m\omega^2 r$ parallel to the axes, its components are $m\omega^2 x$ and $m\omega^2 y$, and the general equation of fluid equilibrium becomes

$$dp=\rho(\omega^2 x dx+\omega^2 y dy-g dz),$$

and therefore $$p=\rho\{\tfrac{1}{2}\omega^2(x^2+y^2)-gz\}+C.$$

The surfaces of equal pressure are therefore paraboloids of revolution, and if the vessel be open at the top, the free surface is given by the equation

$$\omega^2(x^2+y^2)-2gz+\frac{2C}{\rho}=\frac{2\Pi}{\rho},$$

where Π is the external pressure.

The constant must be determined by help of the data of each particular case.

For instance, let the vessel be closed at the top and be just filled with liquid, and let $\Pi=0$; then, taking the origin at the highest point of the axis, $p=0$ when x, y, and z vanish, and therefore $C=0$, and

$$p=\rho\{\tfrac{1}{2}\omega^2(x^2+y^2)-gz\}.$$

27. Next consider the case of elastic fluid enclosed in a vessel which rotates about a vertical axis;

as before $$dp=\rho\{\omega^2(xdx+ydy)-gdz\},$$

and $$p=k\rho;$$

$$\therefore k\log\rho=\tfrac{1}{2}\omega^2(x^2+y^2)-gz+C,$$

so that the surfaces of equal pressure and density are paraboloids.

Let the containing vessel be a cylinder rotating about its axis, and suppose the whole mass of fluid given; then, to determine the constant, consider the fluid arranged in elementary horizontal rings each of uniform density: let r be the radius of one of these rings at a height z, δr its horizontal and δz its vertical thickness, h the height, and a the radius of the cylinder:

$$\text{the mass of the ring}=2\pi\rho r\delta r\delta z,$$

$$\text{and the whole mass } (M) \text{ of the fluid}=\int_0^h\int_0^a 2\pi\rho rdzdr,$$

the origin being taken at the base of the cylinder.

Now $$\rho=e^{\frac{C}{k}}\cdot e^{\frac{\omega^2r^2-2gz}{2k}};$$

and $$\therefore M=\frac{2\pi k^2}{g\omega^2}e^{\frac{C}{k}}\left(e^{\frac{\omega^2a^2}{2k}}-1\right)\left(1-e^{-\frac{gh}{k}}\right),$$

an equation by which C is determined.

28. In general the equation of equilibrium for a fluid revolving uniformly and acted upon by forces of any kind is

$$dp=\rho\{Xdx+Ydy+Zdz+\omega^2(xdx+ydy)\}.$$

In order that the equilibrium may be possible, three equations of condition must be satisfied, expressing that dp is a perfect differential, and, if these conditions are satisfied, the surfaces of equal pressure, and, in certain cases, the free surface can be determined; but it must be observed that a free surface is not always possible. In fact, in order that there may be a free surface, the surfaces of equal pressure must be symmetrical with respect to the axis of rotation.

EXAMPLES

1. A closed tube in the form of an ellipse with its major axis vertical is filled with three different liquids of densities ϱ_1, ϱ_2, ϱ_3 respectively. If the distances of the surfaces of separation from either focus be r_1, r_2, r_3 respectively, prove that

$$r_1(\varrho_2-\varrho_3)+r_2(\varrho_3-\varrho_1)+r_3(\varrho_1-\varrho_2)=0.$$

2. Find the surfaces of equal pressure when the forces tend to fixed centres and vary as the distances from those centres.

3. Prove that if the forces per unit of mass at x, y, z parallel to the axes are

$$y(a-z), \quad x(a-z), \quad xy,$$

the surfaces of equal pressure are hyperbolic paraboloids and the curves of equal pressure and density are rectangular hyperbolas.

4. In a solid sphere two spherical cavities, whose radii are equal to half the radius of the solid sphere, are filled with liquid; the solid and liquid particles attract each other with forces which vary as the distance: prove that the surfaces of equal pressure are spheres concentric with the solid sphere.

5. Show that the forces represented by

$$X=\mu(y^2+yz+z^2), \quad Y=\mu(z^2+zx+x^2), \quad Z=\mu(x^2+xy+y^2)$$

will keep a mass of liquid at rest, if the density $\propto \frac{1}{(\text{dist.})^2}$ from the plane $x+y+z=0$; and the curves of equal pressure and density will be circles.

6. If a conical cup be filled with liquid, the mean pressure at a point in the volume of the liquid is to the mean pressure at a point in the surface of the cup as 3 : 4.

7. A mass of fluid rests upon a plane subject to a central attractive force μ/r^2, situated at a distance c from the plane on the side opposite to that on which the fluid is; and a is the radius of the free spherical surface of the fluid: show that the pressure on the plane

$$=\pi\varrho\mu(a-c)^2/a.$$

8. Find the surfaces of equal pressure for homogeneous fluid acted upon by two forces which vary as the inverse square of the distance from two fixed points.

Prove that if the surface of no pressure be a sphere, the loci of points at which the pressure varies inversely as the distance from one of the centres of force are also spheres.

9. If the components parallel to the axes of the forces acting on an element of fluid at (x, y, z) be proportional to

$$y^2+2\lambda yz+z^2, \quad z^2+2\mu zx+x^2, \quad x^2+2\nu xy+y^2,$$

show that if equilibrium be possible we must have

$$2\lambda=2\mu=2\nu=1.$$

10. A fluid is in equilibrium under a given system of forces; if $\varrho_1=\phi(x, y, z)$ $\varrho_2=\psi(x, y, z)$ be two possible values of the density at any point, show that the equations of the surfaces of equal pressure in either case are given by

$$\phi(x, y, z)+\lambda\psi(x, y, z)=0,$$

where λ is an arbitrary parameter.

11. A hollow sphere of radius a, just full of homogeneous liquid of unit density, is placed between two external centres of attractive force μ^2/r^2 and μ'^2/r'^2, distant c apart, in such a position that the attractions due to them at the centre are equal and opposite. Prove that the pressure at any point is

$$\mu^2/r+\mu'^2/r'-\mu^{\frac{1}{2}}\mu'^{\frac{1}{2}}(\mu+\mu')^2/\{(\mu+\mu')^2a^2+\mu\mu'c^2\}^{\frac{1}{2}}.$$

12. The density of a liquid, contained in a cylindrical vessel, varies as the depth; it is transferred to another vessel, in which the density varies as the square of the depth; find the shape of the new vessel.

13. A rigid spherical shell is filled with homogeneous inelastic fluid, every particle of which attracts every other with a force varying inversely as the square of the distance; show that the difference between the pressures at the surface and at any point within the fluid varies as the area of the least section of the sphere through the point.

14. An open vessel containing liquid is made to revolve about a vertical axis with uniform angular velocity. Find the form of the vessel and its dimensions that it may be just emptied.

15. An infinite mass of homogeneous fluid surrounds a closed surface and is attracted to a point (O) within the surface with a force which varies inversely as the cube of the distance. If the pressure on any element of the surface about a point P be resolved along PO, prove that the whole radial pressure, thus estimated, is constant, whatever be the shape and size of the surface, it being given that the pressure of the fluid vanishes at an infinite distance from the point O.

16. All space being supposed filled with an elastic fluid the particles of which are attracted to a given point by a force varying as the distance, and the whole mass of the fluid being given, find the pressure on a circular disc placed with its centre at the centre of force.

17. A mass m of elastic fluid is rotating about an axis with uniform angular velocity ω, and is acted on by an attraction towards a point in that axis equal to μ times the distance, μ being greater than ω^2; prove that the equation of a surface of equal density ϱ is

$$\mu(x^2+y^2+z^2)-\omega^2(x^2+y^2)=k\log\left\{\frac{\mu(\mu-\omega^2)^2}{8\pi^3}\cdot\frac{m^2}{\varrho^2k^3}\right\}.$$

18. A mass of self-attracting liquid, of density ϱ, is in equilibrium, the law of attraction being that of the inverse square: prove that the mean pressure throughout any sphere of the liquid, of radius r, is less by $\frac{2}{5}\pi\varrho^2r^2$ than the pressure at its centre.

19. A fluid is slightly compressible according to the law

$$(\varrho-\varrho_0)/\varrho_0=\beta(p-p_0)/p_0,$$

where β is small: prove that a mass $\frac{4}{3}\pi\varrho_0a^3$ of the fluid will, under the action of its own gravitation with an external pressure p_0, assume a spherical form of approximate radius $a(1-\frac{4}{45}\beta m\pi a^2\varrho_0^2/p_0)$, where m is the constant of gravitation.

20. A mass M of gas at uniform temperature is diffused through all space, and at each point (x, y, z) the components of force per unit mass are $-Ax$, $-By$, $-Cz$. The pressure and density at the origin are p_0 and ϱ_0 respectively. Prove that

$$ABC\varrho_0M^2=8\pi^3p_0^3.$$

21. A given mass of air is contained within a closed air-tight cylinder with its axis vertical. The air is rotating in relative equilibrium about the axis of the cylinder. The pressure at the highest point of its axis is P, and the

pressure at the highest points of its curved surface is p. Prove that, if the fluid were absolutely at rest, the pressure at the upper end of the axis would be $(p-P)/\{\log p-\log P\}$; where the weight of the air is taken into account.

22. A mass of gas at constant temperature is at rest under the action of forces of potential ψ at any point of space, with any boundary conditions. At the point where ψ is zero, the pressure is Π and the density ϱ_0. The gas is now removed from the action of the forces and confined in a space so that it is at a uniform density ϱ_0. Prove that the loss of intrinsic potential energy by the gas, due to the expansion, is

$$\varrho_0\iiint\psi e^{\frac{-\rho_0\psi}{\Pi}}\,dv;$$

where the integrations are taken throughout the gas in its original state.

23. A uniform spherical mass of liquid of density $\varrho+\sigma$ and radius a is surrounded by another incompressible liquid of density ϱ and external radius b. The whole is in equilibrium under its own gravitation, but with no external pressure. Show that the pressure at the centre is

$$\frac{2}{3}\pi(\varrho+\sigma)^2a^2+\frac{2}{3}\pi\varrho\left\{\frac{2a^2}{b}\sigma+\varrho(a+b)\right\}(b-a).$$

24. A uniform spherical mass of incompressible fluid, of density ϱ and radius a, is surrounded by another incompressible fluid, of density σ and external radius b. The total fluid is in equilibrium under its gravitation, but with no external pressure or forces. The two fluids are now mixed into a homogeneous fluid of the same volume, and the mass is again in equilibrium in a spherical form. Prove that the pressure at the centre in the first case exceeds the pressure at the centre in the second case by

$$\frac{8}{3}\pi\sigma(\varrho-\sigma)a^2\left(1-\frac{a}{b}\right)\left[1+\frac{1}{4}\left(\frac{\varrho}{\sigma}-1\right)\left(1+\frac{a}{b}\right)\left(1+\frac{a^2}{b^2}\right)\right].$$

25. The boundary of a homogeneous gravitating solid, of density σ and mass M, is the surface $r=a\{1+\alpha P_n(\cos\theta)\}$, where α is a quantity so small that its square may be neglected. The solid is surrounded by a mass M' of gravitating liquid, of density ϱ. Show that the equation to the free surface is approximately

$$r=b\{1+\beta P_n(\cos\theta)\},$$

where
$$b^3=\frac{3}{4\pi}\left(\frac{M'}{\varrho}+\frac{M}{\sigma}\right),$$

and
$$\beta=3(\sigma-\varrho)a^{n+3}\alpha/\{(2n-2)\varrho b^3+(2n+1)(\sigma-\varrho)a^3\}b^n.$$

26. A uniform incompressible fluid is of mass M in gravitational units, and forms a sphere of radius a when undisturbed under the influence of its own attraction. It is placed in a weak field of force of gravitational potential

$$\Sigma\mu_n\frac{r^n}{a^{n+1}}P_n(\cos\theta),\ (n>1),$$

where r is measured from the centre of the mean spherical surface of the liquid, and the squares of quantities of the type μ_n can be neglected. Prove that the equation of the free surface is

$$\frac{r}{a}=1+\Sigma\frac{\mu_n}{M}\,\frac{2n+1}{2n-2}P_n(\cos\theta).$$

27. Prove that the pressure at the centre of the Earth, if it were a homogeneous liquid, would be $\frac{1}{3}\varrho a$ lb. per square foot, where ϱ is the mass in pounds

of a cubic foot of the substance of the Earth and a is the Earth's radius in feet.

28. The density of a gravitating liquid sphere of radius a at any point increases uniformly as the point approaches the centre. The surface density is ϱ_0 and the mean density is ϱ. Prove that the pressure at the centre is

$$\tfrac{2}{9}\pi a^2\{10\varrho(\varrho-\varrho_0)+3\varrho_0^2\}.$$

29. In a gravitating fluid sphere of radius a the surfaces of equal density are spheres concentric with the boundary, and the density increases from surface to centre according to any law. Prove that the pressure at the centre is greater than it would be if the density were uniform by

$$\tfrac{8}{9}\pi\gamma\int_0^a(\varrho'^2-\varrho^2)r\,dr,$$

where ϱ denotes the mean density of the whole mass, ϱ' the mean density of that portion which is within a distance r of the centre, and γ is the constant of gravitation.

CHAPTER III

THE RESULTANT PRESSURE OF FLUIDS ON SURFACES

29. In the preceding Chapter we have shown how to investigate the pressure *at any point* of a fluid at rest under the action of given forces; we now proceed to determine the resultants of the pressures exerted by fluids *upon surfaces* with which they are in contact.

We shall consider, first, the action of fluids on plane surfaces, secondly, of fluids under the action of gravity upon curved surfaces, and thirdly, of fluids at rest under any given forces upon curved surfaces.

Fluid Pressures on Plane Surfaces

The pressures at all points of a plane being perpendicular to it, and in the same direction, the resultant pressure is equal to the sum of these pressures.

Hence, if the fluid be incompressible and acted upon by gravity only, the resultant pressure on a plane

$$=\Sigma g\rho z dA,$$

where z is the depth of a small element dA of the area of the plane

$$=g\rho\bar{z}A,$$

where A is the whole area and $\bar{z}$ the depth of its centroid.

In general, if the fluid be of any kind, and at rest under the action of any given forces, take the axes of x and y in the plane, and let p be the pressure at the point (x, y).

The pressure on an element of area $\delta x\delta y = p\delta x\delta y$:

$$\therefore \text{ the resultant pressure} = \iint p\,dx\,dy,$$

the integration extending over the whole of the area considered.

If polar co-ordinates be used, the resultant pressure is given by the expression

$$\iint pr\,dr\,d\theta.$$

30. DEF. *The* **centre of pressure** *is the point at which the direction of the single force, which is equivalent to the fluid pressures on the plane surface, meets the surface.*

The centre of pressure is here defined with respect to plane surfaces only; it will be seen afterwards that the resultant action of fluid on a curved surface is not always reducible to a single force.

In the case of a heavy fluid, it is clear that the centre of pressure of a horizontal area, the pressure on every point of which is the same, is its centroid; and, since pressure increases with the depth, the centre of pressure of any plane area, not horizontal, is below its centroid.

To obtain formulæ for the determination of the centre of pressure of any plane area.

Let p be the pressure at the point (x, y), referred to rectangular axes in the plane, $x+\delta x$, $y+\delta y$, the co-ordinates of an adjacent point,

$\bar{x}$, $\bar{y}$, co-ordinates of the centre of pressure.

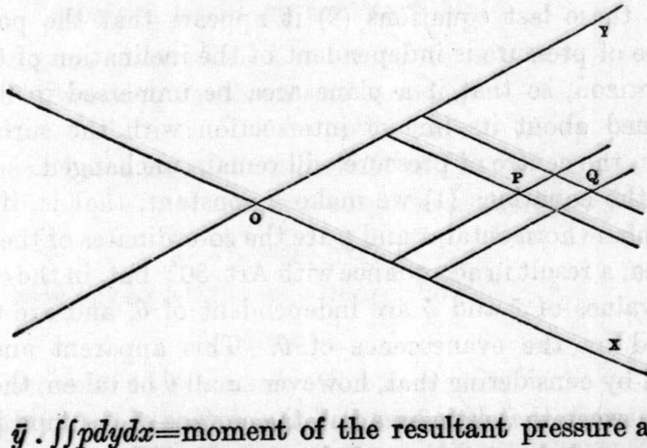

Then $\bar{y} . \iint p dy dx$ = moment of the resultant pressure about OX

= the sum of the moments of the pressures on all the elements of area about OX

$$= \Sigma p \delta y \delta x . y$$

$$= \iint p y dy dx;$$

$$\therefore \bar{y} = \frac{\iint p y dy dx}{\iint p dy dx},$$

and similarly $$\bar{x} = \frac{\iint p x dy dx}{\iint p dy dx},$$

the integrals being taken so as to include the area considered.

If polar co-ordinates be employed, a similar process will give the equations

$$\bar{x}=\frac{\iint pr^2 \cos\theta dr d\theta}{\iint pr dr d\theta}, \quad \bar{y}=\frac{\iint pr^2 \sin\theta dr d\theta}{\iint pr dr d\theta}.$$

31. If the fluid be homogeneous and inelastic, and if gravity be the only force in action,

$$p=g\rho h,$$

where h is the depth of the point P below the surface; and we obtain

$$\bar{x}=\frac{\iint hx dy dx}{\iint h dy dx}, \quad \bar{y}=\frac{\iint hy dy dx}{\iint h dy dx} \quad . \quad . \quad . \quad (1)$$

It is sometimes useful to take for one of the axes the line of intersection of the plane with the surface of the fluid: if we take this line for the axis of x, and θ as the inclination of the plane to the horizon, $p=g\rho y \sin\theta$, and therefore

$$\bar{x}=\frac{\iint xy dy dx}{\iint y dy dx}, \quad \bar{y}=\frac{\iint y^2 dy dx}{\iint y dy dx} \quad . \quad . \quad . \quad (2)$$

From these last equations (2) it appears that the position of the centre of pressure is independent of the inclination of the plane to the horizon, so that if a plane area be immersed in fluid, and then turned about its line of intersection with the surface as a fixed axis, the centre of pressure will remain unchanged.

If in the equations (1) we make h constant, that is, if we suppose the plane horizontal, $\bar{x}$ and $\bar{y}$ are the co-ordinates of the centroid of the area, a result in accordance with Art. 30; but, in the equations (2), the values of $\bar{x}$ and $\bar{y}$ are independent of θ, and are therefore unaffected by the evanescence of θ. This apparent anomaly is explained by considering that, however small θ be taken, the portion of fluid between the plane area and the surface of the fluid is always wedge-like in form, and the pressures at the different points of the plane, although they all vanish in the limit, do not vanish in ratios of equality, but in the constant ratios which they bear to one another for any finite value of θ.

The equations of this article may also be obtained by the following reasoning.

Through the boundary line of the plane area draw vertical lines to the surface enclosing a mass of fluid; then the reaction of the plane, resolved vertically, is equal to the weight of the fluid, which acts in a vertical line through its centre of mass; and the point in which this line meets the plane is the centre of pressure.

Taking the same axes, the weight of an elementary prism, acting through the point (x, y), is $g\rho h\delta x\delta y \cos\theta$, where θ is the inclination of the plane to the horizon; and therefore the centre of these parallel forces acting at points of the plane is given by the equations

$$\bar{x}=\frac{\iint g\rho hx\cos\theta dydx}{\iint g\rho h\cos\theta dydx},\quad \bar{y}=\frac{\iint g\rho hy\cos\theta dydx}{\iint g\rho h\cos\theta dydx},$$

or

$$\bar{x}=\frac{\iint hxdydx}{\iint hdydx},\qquad \bar{y}=\frac{\iint hydydx}{\iint hdydx}.$$

Hence it appears that the depth of the centre of pressure is double that of the centre of mass of the fluid enclosed.

32. The following theorem determines geometrically the position of the centre of pressure for the case of a heavy liquid.

If a straight line be taken in the plane of the area, parallel to the surface of the liquid and as far below the centroid of the area as the surface of the liquid is above, the pole of this straight line with respect to the momental ellipse at the centroid whose semi-axes are equal to the principal radii of gyration at that point will be the centre of pressure of the area.

Taking A for the area, and b, a for the principal radii of gyration, these are determined by the equations

$$Ab^2=\iint y^2dxdy,\quad Aa^2=\iint x^2dxdy,$$

and the equation of the momental ellipse is

$$\frac{x^2}{a^2}+\frac{y^2}{b^2}=1,$$

the co-ordinate axes being the principal axes at the centroid.

Let $\bar{x}$, $\bar{y}$ be the co-ordinates of the centre of pressure, and

$$x\cos\theta+y\sin\theta=p$$

the equation to the line in the surface;

then

$$\bar{x}=\frac{\iint(p-x\cos\theta-y\sin\theta)xdxdy}{\iint(p-x\cos\theta-y\sin\theta)dxdy}=-\frac{a^2}{p}\cos\theta,$$

and similarly,

$$\bar{y}=-\frac{b^2}{p}\sin\theta;$$

$\therefore$ $(\bar{x}, \bar{y})$ is the pole of the line

$$x\cos\theta+y\sin\theta=-p$$

with respect to the momental ellipse.

33. *Examples of the determination of centres of pressure.*

(1) *A quadrant of a circle just immersed vertically in a heavy homogeneous liquid, with one edge in the surface.*

If Ox, the edge in the surface, be the axis of x,

$$\bar{x}=\frac{\int_0^a\int_0^{\sqrt{(a^2-x^2)}} xydxdy}{\int_0^a\int_0^{\sqrt{(a^2-x^2)}} ydxdy},\quad \bar{y}=\frac{\iint y^2dxdy}{\iint ydxdy},$$

the limits of the integrations for $\bar{y}$ being the same as for $\bar{x}$.

$$\iint ydxdy=\tfrac{1}{2}\int(a^2-x^2)dx=\tfrac{1}{3}a^3,$$

$$\iint xydxdy=\tfrac{1}{2}\int x\,.\,(a^2-x^2)dx=\tfrac{1}{8}a^4,$$

$$\iint y^2dxdy=\tfrac{1}{3}\int(a^2-x^2)^{\frac{3}{2}}dx=\frac{\pi a^4}{16};$$

$$\therefore\ \bar{x}=\frac{3}{8}a,\ \bar{y}=\frac{3}{16}\pi a.$$

Employing polar co-ordinates and taking the line Ox as the initial line, we should have $p=g\rho r\sin\theta$, and

$$\bar{x}=\frac{\int_0^a\int_0^{\frac{\pi}{2}} r^3\cos\theta\sin\theta drd\theta}{\iint r^2\sin\theta drd\theta}=\frac{3}{8}a,\ \text{and}\ \ \bar{y}=\frac{\int_0^a\int_0^{\frac{\pi}{2}} r^3\sin^2\theta drd\theta}{\iint r^2\sin\theta drd\theta}=\frac{3}{16}\pi a.$$

(2) *A circular area, radius* a, *is immersed with its plane vertical, and its centre at a depth* c.

Take the centre as the origin, and the vertical downwards from the centre as the initial line; then if p be the pressure at the point (r, θ),

$$p=g\rho(c+r\cos\theta),$$

and the depth below the centre of the centre of pressure

$$=\frac{2\int_0^a\int_0^{\pi} r^2\cos\theta(c+r\cos\theta)drd\theta}{2\iint r(c+r\cos\theta)drd\theta}=\frac{a^2}{4c}.$$

It will be seen that this result is at once given by the theorem of Art. 32.

(3) *A vertical rectangle, exposed to the action of the atmosphere at a constant temperature.*

If Π be the atmospheric pressure at the base of the rectangle, the pressure at a height z is $\Pi e^{-\frac{gz}{k}}$, Art. 24, and if b denote the breadth, the pressure upon a horizontal strip of the rectangle

$$=\Pi e^{-\frac{gz}{k}}b\delta z,$$

$\therefore$ the resultant pressure, if a be the height,

$$=\int_0^a \Pi e^{-\frac{gz}{k}}bdz=\Pi\frac{bk}{g}\left(1-e^{-\frac{ga}{k}}\right),$$

and the height of the centre of pressure

$$=\frac{\int_0^a ze^{-\frac{gz}{k}}dz}{\int_0^a e^{-\frac{gz}{k}}dz}=\frac{k}{g}-\frac{a}{e^{\frac{ga}{k}}-1}.$$

(4) *A hollow cube is very nearly filled with liquid, and rotates uniformly about a diagonal which is vertical; required to find the pressures upon, and the centres of pressure of, its several faces.*

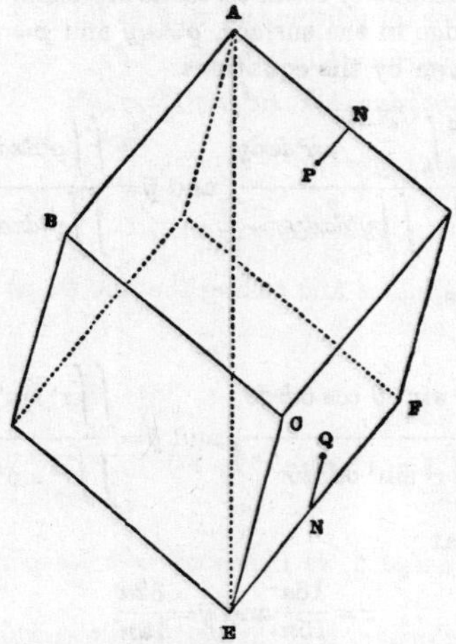

I. For one of the upper faces *ABCD*, take *AD*, *AB*, as axes of x and y; z, r, the vertical and horizontal distances of any point P (x, y) from A,

then $\qquad \frac{p}{\varrho}=\frac{1}{2}\omega^2r^2+gz,$

$$z=\frac{x+y}{\sqrt{3}}, \text{ projecting the broken line } ANP \text{ on } AE,$$

$$r^2=AP^2-z^2=x^2+y^2-z^2=\tfrac{2}{3}(x^2+y^2-xy);$$

$\therefore$ the pressure (P) on $ABCD=\int_0^a\int_0^a pdydx$

$$=\varrho\,.\iint\left\{\frac{\omega^2}{3}(x^2+y^2-xy)+\frac{g}{\sqrt{3}}(x+y)\right\}dydx$$

$$=\varrho\left\{\frac{5}{36}a^4\omega^2+\frac{g}{\sqrt{3}}a^3\right\}.$$

The centre of pressure is given by the equations

$$\bar{x}P=\bar{y}P=\varrho\int_0^a\int_0^a x\left\{\frac{\omega^2}{3}(x^2+y^2-xy)+\frac{g}{\sqrt{3}}(x+y)\right\}dydx$$

$$\therefore \bar{x}=\bar{y}=a\cdot\frac{21g+3\sqrt{3}\omega^2a}{36g+5\sqrt{3}\omega^2a}.$$

II. For one of the lower faces $ECDF$, take EF, EC as axes, then, for a point Q,

$$z=a\sqrt{3}-\frac{x+y}{\sqrt{3}},$$

$$r^2=\tfrac{2}{3}(x^2+y^2-xy),$$

and the rest of the process is the same as in the first case.

(5) *A quadrant of a circle is just immersed vertically, with one edge in the surface, in a liquid, the density of which varies as the depth.*

Taking Ox as the edge in the surface, $\rho=\mu y$ and $p=\frac{1}{2}\mu g y^2$; the centre of pressure is therefore given by the equations

$$\bar{x}=\frac{\int_0^a\int_0^{\sqrt{a^2-x^2}} xy^2dxdy}{\iint y^2dxdy}, \text{ and } \bar{y}=\frac{\iint y^3dxdy}{\iint y^2dxdy};$$

or, in polar co-ordinates,

$$x=\frac{\int_0^a\int_0^{\frac{\pi}{2}} r^4 \sin^2\theta\cos\theta drd\theta}{\iint r^3\sin^2\theta drd\theta}, \text{ and } \bar{y}=\frac{\iint r^4\sin^3\theta drd\theta}{\iint r^3\sin^2\theta drd\theta};$$

and it will be found that

$$\bar{x}=\frac{16a}{15\pi} \text{ and } \bar{y}=\frac{32a}{15\pi}.$$

34. Let G be the centre of gravity and C the centre of pressure of a plane area A which is moved parallel to itself so that the depth of G is increased from $\bar{z}$ to $\bar{z}+h$, and let G', C' be the new positions of G, C. Then the pressure at every point of A is increased by the same amount $g\rho h$ and the resultant pressure is therefore increased by adding a force $g\rho hA$, acting at G', to the original resultant $g\rho\bar{z}A$ which acts at C', so that the new centre of pressure C'' is on $G'C'$ and divides it so that

$$G'C'' : G'C' = \bar{z} : \bar{z}+h.$$

If a given plane area turn in its own plane about a fixed point, the centre of pressure changes its position and describes a curve on the area.

If AB is the line of intersection of the plane area with the surface, the distance of the centre of pressure from AB is independent of the inclination of the area to the vertical (Art. 31).

We may therefore take the area to be vertical.

Let h be the depth of the fixed point O, and let Ox, Oy be axes fixed in the area.

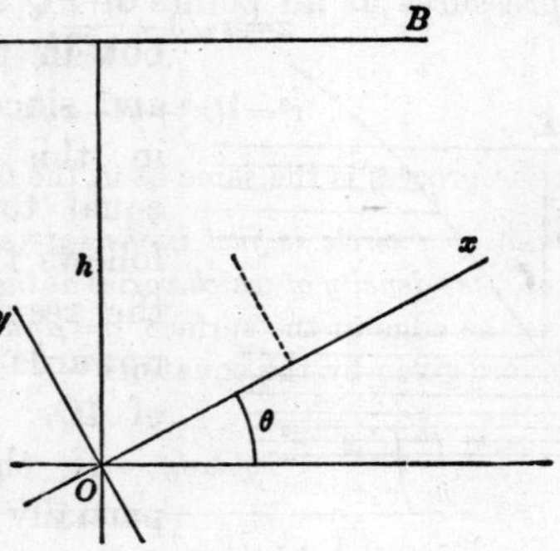

Then, if θ is the inclination of Ox to the horizontal,

$$p = g\rho(h - x\sin\theta - y\cos\theta).$$

$$\therefore \bar{x} = \frac{\iint pxdxdy}{\iint pdxdy} = \frac{a + b\sin\theta + c\cos\theta}{d + e\sin\theta + f\cos\theta},$$

and
$$\bar{y} = \frac{a' + b'\sin\theta + c'\cos\theta}{d + e\sin\theta + f\cos\theta},$$

a, b, d, etc., being known constants, and the elimination of θ gives a conic section as the locus of the centre of pressure.

Resultant Pressures on Curved Surfaces

35. *To find the resultant vertical pressure on any surface of a homogeneous liquid at rest under the action of gravity.*

PQ being a surface exposed to the action of a heavy liquid, let AB be the projection of PQ on the surface of the liquid.

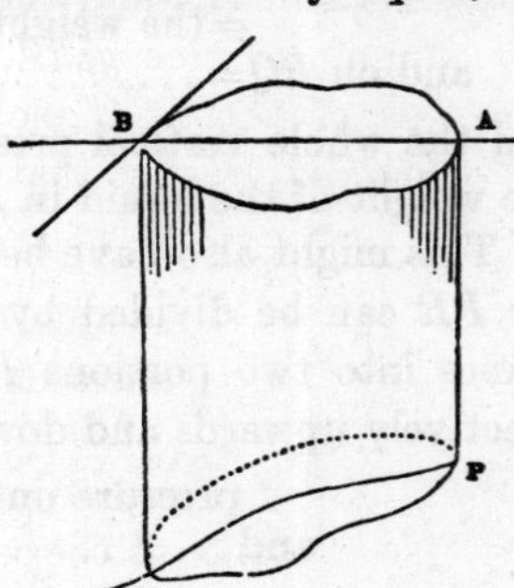

The mass AQ is supported by the horizontal pressure of the liquid and by the reaction of PQ; this reaction resolved vertically must be equal to the weight of AQ, and conversely, the vertical pressure on PQ is equal to the weight of AQ, and acts through its centre of mass.

If PQ be pressed upwards by the liquid as in the next figure, produce the surface, project PQ on it as before, suppose the space AQ to be filled with liquid of the same kind, and remove the liquid from the inside.

Then the pressures at all points of PQ are the same as before, but in the contrary direction, and since the vertical pressure in this hypothetical case is equal to the weight of AQ, it follows that in the actual case, the resultant vertical pressure upwards is equal to the weight of AQ.

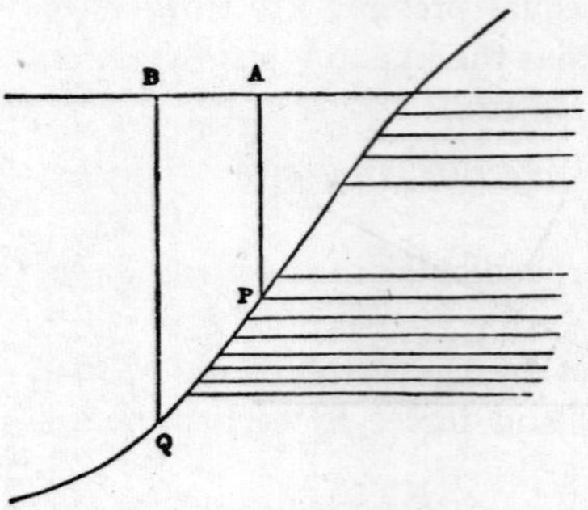

If the surface be pressed partially upwards and partially downwards, draw through P, the highest point of the portion of surface considered, a vertical plane PR, and let ACB be the projection of PSQ on the surface of the liquid.

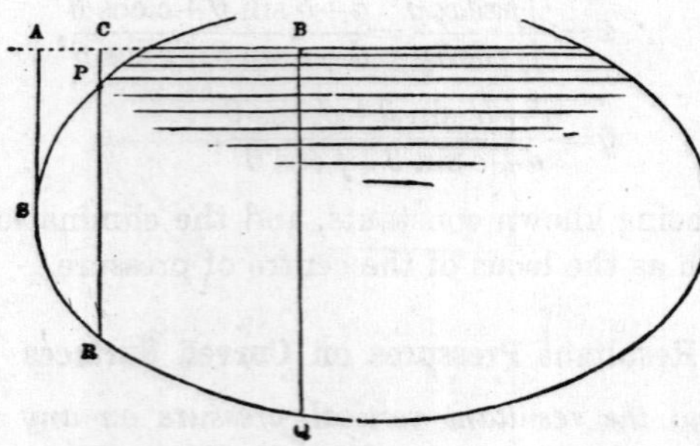

Then the resultant vertical pressure on PSR

=the weight of the liquid in PSR,

and on RQ=.........................CQ,

and the whole vertical pressure=the weight of the liquid in CQ+ the weight of the liquid in PSR.

This might also have been deduced from the two previous cases, for PR can be divided by the line of contact of vertical tangent planes into two portions PS, SR, on which the pressures are respectively upwards and downwards ; and since

pressure on PS=weight of liquid APS,

and..........SR=.............. ASR,

the difference of these, *i.e.* the vertical pressure on PSR=weight of fluid PSR.

In a similar manner other cases may be discussed.

It will be observed that this investigation applies also to the case of a heterogeneous liquid (in which the density must be a function of the depth, since surfaces of equal pressure are surfaces of equal density), provided we consider that the hypothetical extension of the liquid follows the same law of density.

36. *To find the resultant horizontal pressure, in a given direction, on a surface* PQ.

Project PQ on a vertical plane perpendicular to the given direction, and let pq be the projection.

Then the mass Pq is kept at rest by the pressure on pq, the resultant horizontal pressure on PQ, and forces in vertical planes parallel to the plane pq.

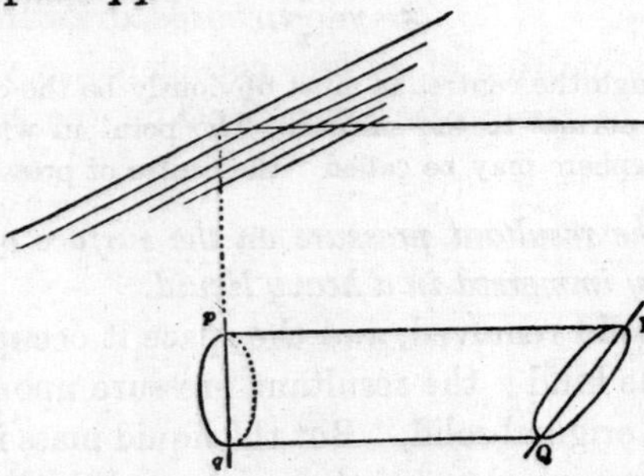

Hence the horizontal pressure on PQ is equal to that on pq, and acts in the same straight line, *i.e.* through the centre of pressure of pq.

Hence, in general, to determine the resultant fluid pressure on any surface, find the vertical pressure, and the resultant horizontal pressures in two directions at right angles to each other. These three forces may in some cases be compounded into a single force, the condition for which may be determined by the usual methods of Statics.

EXAMPLE. *A hemisphere is filled with homogeneous liquid: required to find the resultant action on one of the four portions into which it is divided by two vertical planes through its centre at right angles to each other.*

Taking the centre O as origin, the bounding horizontal radii as axes of x and y, and the vertical radius as the axis of z, the pressure parallel to x is equal to the pressure on the quadrant yOz, which is the projection, on a plane perpendicular to Ox, of the curved surface.

Therefore, the pressure parallel to Ox

$$=g\rho\frac{\pi a^2}{4}\cdot\frac{4a}{3\pi}=\frac{1}{3}g\rho a^3,$$

and the co-ordinates of its point of action are

$$(0, \tfrac{3}{8}a, \tfrac{3}{16}\pi a), \text{ Art 33, Ex. 1};$$

similarly, the pressure parallel to $Oy = \frac{1}{8}g\varrho a^3$, and acts through the point

$$(\tfrac{3}{8}a, 0, \tfrac{3}{16}\pi a).$$

The resultant vertical pressure=the weight of the liquid $= \frac{1}{6}g\varrho\pi a^3$, and acts in the direction of the line $x=y=\frac{3}{8}a$.

The directions of the three forces all pass through the point

$$(\tfrac{3}{8}a, \tfrac{3}{8}a, \tfrac{3}{16}\pi a),$$

and they are therefore equivalent to a single force

$$\tfrac{1}{8}g\varrho a^3\sqrt{(\pi^2+8)}$$

in the line

$$x-\frac{3}{8}a=y-\frac{3}{8}a=\frac{2}{\pi}\left(z-\frac{3}{16}\pi a\right),$$

or

$$x=y=\frac{2}{\pi}z,$$

a straight line through the centre, as must obviously be the case, since all the fluid pressures are normal to the surface. The point in which it meets the surface of the hemisphere may be called "the centre of pressure."

37. *To find the resultant pressure on the surface of a solid either wholly or partially immersed in a heavy liquid.*

Suppose the solid removed, and the space it occupied filled with liquid of the same kind; the resultant pressure upon it will be the same as upon the original solid. But the liquid mass is at rest under the action of its own weight, and the pressure of the liquid surrounding it: the resultant pressure is therefore equal to the weight of the liquid displaced, and acts in a vertical line through its centre of mass.

The same reasoning evidently shows that the resultant pressure of an elastic fluid on any solid is equal to the weight of the elastic fluid displaced by the solid.

This result may also be obtained by means of Arts. 35 and 36, as follows: Draw parallel horizontal lines touching the surface, and forming a cylinder which encloses it; the curve of contact divides the surface into two parts, on which the resultant horizontal pressures, parallel to the axis of the cylinder, are equal and opposite; the horizontal pressures on the solid therefore balance each other and the resultant is wholly vertical. To determine the amount of the resultant vertical pressure, draw parallel vertical lines touching the surface, and dividing it into two portions on one of which the resultant vertical pressure acts upwards, and on the other downwards; the difference of the two is evidently the weight of the fluid displaced by the solid.

38. If a solid of given volume (V) be completely immersed in a heavy liquid, and if the surface of the solid consist partly of a curved surface and partly of known plane areas, the resulting pressure on the curved surface can be determined.

For the plane areas being known in size and position, we can calculate the resultant horizontal and the resultant vertical pressure, X and Y, upon those areas; and, since the resulting pressure on the whole surface is vertical and equal to $g\rho V$ upwards, it follows that the resultant horizontal and vertical pressures on the curved surface are respectively equal to X and $g\rho V - Y$.

EXAMPLE. *A solid is formed by turning a circular area round a tangent line through an angle θ, and this solid is held under water with its lower plane face horizontal and at a given depth* h.

In this case,

$$V = \pi a^2\theta, \quad X = g\varrho\pi a^2(h - a\sin\theta)\sin\theta,$$

and

$$Y = g\varrho\pi a^2(h - h\cos\theta + a\sin\theta\cos\theta).$$

39. *To find the resultant pressure on any surface of a fluid at rest under the action of any given forces.*

Let p be the pressure, determined as in Chapter II, at any point (x, y, z) of a surface, S, exposed to the action of a fluid. Let l, m, n be the direction-cosines of the normal at the point (x, y, z).

Let δS be an element of the surface about the same point. The pressures on this element, parallel to the axes, are

$$lp\delta S, \; mp\delta S, \; np\delta S,$$

$\therefore$ if X, Y, Z, and L, M, N, be the resultant pressures parallel to the axes, and the resultant couples, respectively,

$$X = \iint lpdS, \; Y = \iint mpdS, \; Z = \iint npdS,$$

$$L = \iint p(ny - mz)dS,$$

$$M = \iint p(lz - nx)dS,$$

$$N = \iint p(mx - ly)dS,$$

the integrations being made to include the whole of the surface under consideration.

These resultants are equivalent to a single force if

$$XL + YM + ZN = 0.$$

40. The surface may be divided into elements in three different ways by planes parallel to the co-ordinate planes.

Thus, $\delta x \delta y$=projection of δS on $xy = n\delta S$;

and $\therefore Z = \iint p dx dy$; and similarly, $X = \iint p dy dz$, and $Y = \iint p dz dx$,

$$L = \iint p(y dx dy - z dz dx)$$
$$= \iint p(y dy - z dz) dx,$$
$$M = \iint p(z dz - x dx) dy,$$
$$N = \iint p(x dx - y dy) dz.$$

41. If the fluid be at rest under the action of gravity only, and the axis of z be vertical, p is a function of z, $\phi(z)$ suppose, and therefore

$$X = \iint \phi(z) dy dz,$$

which is evidently the expression for the pressure, parallel to x, upon the projection of the given surface on the plane yz; and similarly Y is equal to the pressure upon the projection on xz.

Again, if the fluid be incompressible and acted upon by gravity only, $p\delta x \delta y$ is equal to the weight of the portion of fluid contained between δS and its projection on the surface of the fluid;

$\therefore Z$, or $\iint p dx dy$, is the weight of the superincumbent fluid.

These results accord with those previously obtained, Arts. 35 and 36.

42. When the surface S is closed, as for example the surface of a solid body, it is sometimes convenient to use Green's Theorem to transform the surface integrals of Art. 39 into volume integrals through the space bounded by S. The forces and couples then become

$$X = \iiint \frac{\partial p}{\partial x} dx dy dz,$$

and two similar equations, and

$$L = \iiint \left(y\frac{\partial p}{\partial z} - z\frac{\partial p}{\partial y} \right) dx dy dz$$

and two similar equations; when p is the value of the pressure function at the point (x, y, z) of the enclosed space supposed to contain fluid with the same law of pressure as the surrounding fluid.

43. If a solid body be wholly or partially immersed in any fluid which is at rest under the action of given forces, the resultant fluid pressure on the body will be equal to the resultant of the forces which would act on the displaced fluid.

For we can imagine the solid removed and the gap filled up with the fluid, which will be in equilibrium under the action of the forces and the pressure of the surrounding fluid; and the resultant pressure must be equal and opposite to the resultant of the forces.

In filling up the gap with fluid, the law of density must be maintained, that is, the surfaces of equal density must be continuous with those of the surrounding fluid.

EXAMPLES

1. A heavy thick rope, the density of which is double the density of water, is suspended by one end, outside the water, so as to be partly immersed; find the tension of the rope at the middle of the immersed portion.

2. A hollow sphere of radius a is just filled with water; find the resultant vertical pressures on the two portions of the surface divided by a plane at depth c below the centre.

3. A vessel in the form of a regular pyramid, whose base is a plane polygon of n sides, is placed with its axis vertical and vertex downwards and is filled with fluid. Each side of the vessel is movable about a hinge at the vertex, and is kept in its place by a string fastened to the middle point of its base and to the centre of the polygon; show that the tension of each string is to the whole weight of the fluid as 1 to $n \sin 2\alpha$, where α is the inclination of each side to the horizon.

4. If an area is bounded by two concentric semicircles with their common bounding diameter in the free surface, prove that the depth of the centre of pressure is

$$\tfrac{3}{16}\pi(a+b)(a^2+b^2)/(a^2+b^2+ab),$$

where a and b are the radii.

5. A square lamina $ABCD$, which is immersed in water, has the side AB in the surface; draw a line BE to a point E in CD such that the pressures on the two portions may be equal. Prove that, if this be the case, the distance between the centres of pressure : the side of the square :: $\sqrt{505}:48$.

6. A semicircular lamina is completely immersed in water with its plane vertical, so that the extremity A of its bounding diameter is in the surface, and the diameter makes with the surface an angle α. Prove that if E be the centre of pressure and θ the angle between AE and the diameter,

$$\tan\theta = \frac{3\pi+16\tan\alpha}{16+15\pi\tan\alpha}.$$

7. A plane area immersed in a fluid moves parallel to itself and with its centre of gravity always in the same vertical straight line. Show (1) that the locus of the centres of pressure is a hyperbola, one asymptote of which is the given vertical, and (2) that if a, $a+h$, $a+h'$, $a+h''$ be the depths of the C.G. in any positions, y, $y+k$, $y+k'$, $y+k''$ those of the centre of pressure in the same positions, then

$$\begin{vmatrix} k, & h, & h(k-h) \\ k', & h', & h'(k'-h') \\ k'', & h'', & h''(k''-h'') \end{vmatrix} = 0.$$

8. A right cone is totally immersed in water, the depth of the centre of its base being given. Prove that, P, P', P'' being the resultant pressures on its convex surface, when the sines of the inclination of its axis to the horizon are s, s', s'' respectively,

$$P^2(s'-s'')+P'^2(s''-s)+P''^2(s-s')=0.$$

9. A quantity of liquid acted upon by a central force varying as the distance is contained between two parallel planes; if A, B be the areas of the planes in contact with the fluid, show that the pressures upon them are in the ratio $A^2 : B^2$.

10. A solid sphere rests on a horizontal plane and is just totally immersed in a liquid. It is then divided by two planes drawn through its vertical diameter perpendicular to each other. Prove that if ϱ be the density of the solid, σ that of the fluid, the parts will not separate provided $\sigma>\frac{1}{4}\varrho$.

11. A closed cylinder, very nearly filled with liquid, rotates uniformly about a generating line, which is vertical; find the resultant pressure on its curved surface.

Determine also the point of action of the pressure on its upper end.

12. Show that the depth of the centre of pressure of the area included between the arc and the asymptote of the curve

$$(r-a)\cos\theta=b \text{ is } \frac{a}{4}\cdot\frac{3\pi a+16b}{3\pi b+4a},$$

the asymptote being in the surface and the plane of the curve vertical.

13. If a plane area immersed in a liquid revolve about any axis in its own plane, prove that the centre of pressure describes a straight line in the plane.

14. A solid is formed by turning a parabolic area, bounded by the latus rectum, about the latus rectum, through an angle θ; and this solid is held under water, just immersed, with its lower plane face horizontal. Prove that, if ϕ be the inclination to the horizon of the resultant pressure on the curved surface of the solid,

$$3\sin^2\theta\tan\phi=5\sin\theta-3\sin\theta\cos\theta-2\theta.$$

15. A given area is immersed vertically in a heavy liquid and a cone is constructed on it as base, the cone being wholly immersed: find the locus of the vertex when the resultant pressure on the curved surface is constant, and show that this pressure is unaltered by turning the cone round the horizontal line drawn through the centre of gravity of the base perpendicular to the plane of the base.

16. A vessel in the form of an elliptic paraboloid, whose axis is vertical, and equation $\frac{x^2}{a^2}+\frac{y^2}{b^2}=\frac{z}{h}$, is divided into four equal compartments by its principal planes. Into one of these water is poured to the depth h; prove that, if the resultant pressure on the curved portion be reduced to two forces, one vertical and the other horizontal, the line of action of the latter will pass through the point $(\frac{5}{16}a, \frac{5}{16}b, \frac{3}{7}h)$.

17. A regular polygon wholly immersed in a liquid is movable about its centre of gravity; prove that the locus of the centre of pressure is a sphere.

18. A hemispherical bowl is filled with water, and two vertical planes are drawn through its central radius, cutting off a semi-lune of the surface; if $2a$

be the angle between the planes, prove that the angle which the resultant pressure on the surface makes with the vertical

$$=\tan^{-1}\left(\frac{\sin\alpha}{\alpha}\right).$$

19. A volume $\frac{4}{3}\pi a^3$ of fluid of density ρ surrounds a fixed sphere of radius b and is attracted to a point at a distance $c(<b)$ from its centre by a force μr per unit mass ; supposing the external pressure zero, find the resultant pressure on the fixed sphere.

20. A vessel in the form of a surface of revolution has the following property ; if it be placed with its axis vertical, and any quantity of water be poured into it, the resultant vertical pressure has a constant ratio to the resultant horizontal pressure on either of the portions into which the surface is divided by a vertical plane through its axis ; find the form of the surface.

21. Find the equation of a curve symmetrical about a vertical axis, such that, when it is immersed with its highest point at half the depth of its lowest, the centre of pressure may bisect the axis.

22. A rectangular area is immersed in compressible liquid with its plane vertical and one side in the surface, where the pressure is zero. Show that, if the density is a linear function of the pressure, the depth of the centre of pressure is

$$\frac{a}{m}\,\frac{(m-1)\rho_1+(1-\frac{1}{2}m^2)\rho_0}{\rho_1-(m+1)\rho_0},$$

where a is the length of the vertical side, ρ_0, ρ_1 are the densities at the top and bottom of the area, and

$$m=\log(\rho_1/\rho_0).$$

23. A cubical box of side a has a heavy lid of weight W movable about one edge. It is filled with water, and held with the diagonal through one extremity of this edge vertical. If it be now made to rotate with uniform angular velocity ω, show that, in order that no water may be spilled, W must not be less than

$$\left(\frac{7}{6}+\frac{1}{2\sqrt{3}}\,\frac{\omega^2 a}{g}\right)W',$$

if W' is the weight of the water in the box.

24. A small solid body is held at rest in a fluid in which the pressure p at any point is a given function of the rectangular co-ordinates x, y, z ; prove that the components of the couple which tends to make it rotate round the centre of gravity of its volume are

$$(C-B)\frac{\partial^2 p}{\partial y\partial z}-D\left(\frac{\partial^2 p}{\partial y^2}-\frac{\partial^2 p}{\partial z^2}\right)-E\frac{\partial^2 p}{\partial y\partial x}+F\frac{\partial^2 p}{\partial z\partial x},$$

and two similar expressions, where A, B, C, D, E, F are the moments and products of inertia of the volume of the solid with respect to axes through the centre of gravity.

25. A mass of homogeneous liquid is at rest under the action of forces whose potential is a quadratic function of rectangular co-ordinates, so that the surfaces of equipressure are ellipsoids. Show that, if a body of any shape is held immersed in the liquid, the resultant thrust on the body may be represented as a force acting through G, the centroid of its volume, and directed along the normal to the surface of equipressure through G, together with a

couple which depends on the orientation of the body but not on the position of G in the liquid.

26. A rigid spherical shell of radius a contains a mass M of gas in which the pressure is κ times the density, and the gas is repelled from a fixed external point O (distant c from the centre) with a force per unit of mass equal to κ/(distance). Prove that the resultant pressure of the gas on the shell is

$$\frac{\kappa M}{c}\frac{5c^2-a^2}{5c^2+a^2}.$$

27. A vessel full of water is in the form of an eighth part of an ellipsoid (axes a, b, c), bounded by the three principal planes. The axis c is vertical, and the atmospheric pressure is neglected. Prove that the resultant fluid pressure on the curved surface is a force of intensity

$$\tfrac{1}{8}g\varrho\{b^2c^4+a^2c^4+\tfrac{1}{4}\pi^2a^2b^2c^2\}^{\frac{1}{2}}.$$

28. A hollow ellipsoid is filled with water and placed with its a-axis making an angle α with the horizontal and its c-axis horizontal. Prove that the fluid pressure on the curved surface on either side of the vertical plane through the a-axis is equivalent to a wrench of pitch

$$\frac{3c\sin\alpha\cos\alpha}{2}\cdot\frac{a^2-b^2}{4c^2+9(a^2\sin^2\alpha+b^2\cos^2\alpha)}.$$

29. The angular points of a triangle immersed in a liquid whose density varies as the depth are at distances α, β, γ respectively below the surface, show that the centre of pressure is at a depth

$$\frac{3}{5}\cdot\frac{(\alpha+\beta+\gamma)(\alpha^2+\beta^2+\gamma^2)+\alpha\beta\gamma}{\alpha^2+\beta^2+\gamma^2+\alpha\beta+\beta\gamma+\gamma\alpha}.$$

30. A plane area, completely submerged in a heavy heterogeneous fluid, rotates about a fixed horizontal axis at depth h perpendicular to its plane. If the density of the fluid at depth z be equal to μz, and if the area be symmetrical about each of two rectangular axes meeting at the point of intersection of the area with the axis of rotation, prove that the locus in space of the centre of pressure is an ellipse with its centre at a depth

$$2h-\frac{h(a^4-k_1^2k_2^2)}{(a^2+k_1^2)(a^2+k_2^2)},$$

where k_1 and k_2 are the radii of gyration of the area with respect to the axes of symmetry and the atmospheric pressure is

$$\tfrac{1}{2}g\mu(a^2-h^2).$$

31. Show that the pressure on any plane area immersed in water can be reduced to a force at the centroid of the area, and a couple about an axis in the plane of the area, and that the axis of this couple is perpendicular to the tangent at the end of the horizontal diameter of a momental ellipse at the centroid.

CHAPTER IV

THE EQUILIBRIUM OF FLOATING BODIES

44. *To find the conditions of equilibrium of a floating body.*

We shall suppose that the fluid is at rest under the action of gravity only, and that the body, under the action of the same force, is floating freely in the fluid. The only forces then which act on the body are its weight, and the pressure of the surrounding fluid, and in order that equilibrium may exist, the resultant fluid pressure must be equal to the weight of the body, and must act in a vertical direction.

Now we have shown that the resultant pressure of a heavy fluid on the surface of a solid, either wholly or partially immersed, is equal to the weight of the fluid displaced, and acts in a vertical line through its centre of mass.

Hence it follows that the weight of the body must be equal to the weight of the fluid displaced, and that the centres of mass of the body, and of the fluid displaced, must lie in the same vertical line.

These conditions are necessary and sufficient conditions of equilibrium, whatever be the nature of the fluid in which the body is floating. If it be heterogeneous, the displaced fluid must be looked upon as following the same law of density as the surrounding fluid; in other words, it must consist of strata of the same kind as, and continuous with, the horizontal strata of uniform density, in which the particles of the surrounding fluid are necessarily arranged.

If for instance a solid body float in water, partially immersed, its weight will be equal to the weight of the water displaced, together with the weight of the air displaced; and if the air be removed, or its pressure diminished by a diminution of its density or temperature, the solid will sink in the water through a space depending upon its own weight, and upon the densities of air and water. This may be further explained by observing that the pressure of the air on the water is greater than at any point above it, and that this surface pressure of the air is transmitted by the water to the immersed portion of the floating body, and consequently the upward pressure of the air upon it is greater than the downward pressure.

45. We now proceed to illustrate the application of the above conditions, by discussion of some particular cases.

EXAMPLE 1. *A portion of a solid paraboloid, of given height, floats with its axis vertical and vertex downwards in a homogeneous liquid: required to find its position of equilibrium.*

Taking $4a$ as the latus rectum of the generating parabola, h its height, and x the depth of its vertex, the volumes of the whole solid and of the portion immersed are respectively $2\pi ah^2$ and $2\pi ax^2$; and if ϱ, σ be the densities of the solid and liquid, one condition of equilibrium is

$$\varrho \cdot 2\pi ah^2 = \sigma \cdot 2\pi ax^2;$$

$$\therefore x = \sqrt{\frac{\varrho}{\sigma}}h,$$

which determines the portion immersed, the other condition being obviously satisfied.

EXAMPLE 2. *It is required to find the positions of equilibrium of a square lamina floating with its plane vertical, in a liquid of double its own density.*

The conditions of equilibrium are clearly satisfied if the lamina float half immersed either with a diagonal vertical, or with two sides vertical.

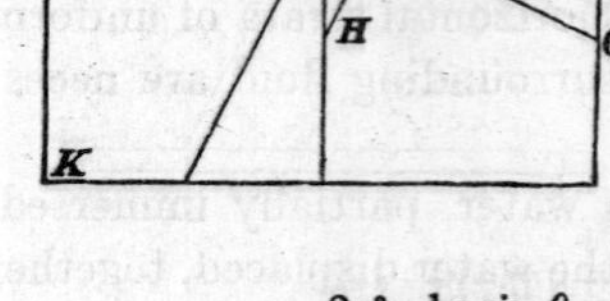

To examine whether there is any other position of equilibrium, let the lamina be held with the line DGC in the surface, in which case the first condition is satisfied.

But, if the angle $CGA = \theta$, and if $2a$ be the side of the square, the moment about G of the fluid pressure, which is the same as the difference between the moments of the rectangle AK, and of twice the triangle GBD, is proportional to

$$2a^2 \cdot \tfrac{1}{2}a \sin\theta - a^2 \tan\theta \cdot \tfrac{1}{3}(a \sec\theta + a\cos\theta),$$

or to $$\sin\theta(1 - \tan^2\theta),$$

and this vanishes only when $\theta = 0$ or $\tfrac{1}{4}\pi$.

Hence there is no other position of equilibrium.

Example 3. *A triangular prism floats with its edges horizontal, to find its positions of equilibrium.*

Let the figure be a section of the prism by a vertical plane through its centre of gravity.

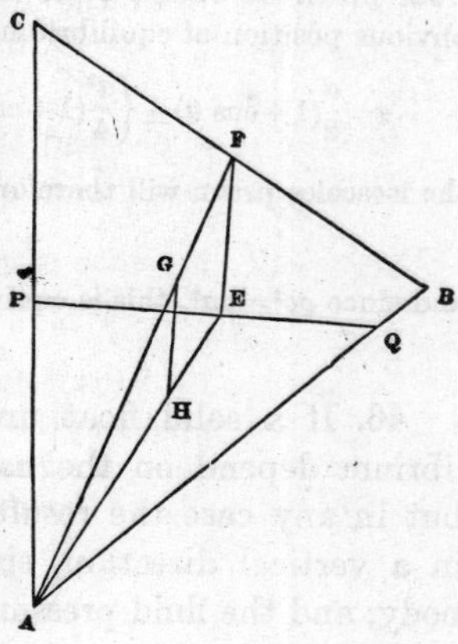

PQ is the line of flotation and H the centre of gravity of the liquid displaced. When there is equilibrium the area APQ is to ABC in the ratio of the density of the prism to the density of the liquid, and therefore for all possible positions of PQ the area APQ is constant; hence PQ always touches, at its middle point, an hyperbola of which AB, AC are the asymptotes.

Also HG must be perpendicular to PQ, and therefore since

$$AH : HE = AG : GF,$$

FE must be perpendicular to PQ, that is, FE is the normal at E to the hyperbola. The problem is therefore reduced to that of drawing normals from F to the curve.

Let
$$xy = c^2 \quad . \quad . \quad . \quad . \quad . \quad (1)$$
be the equation of the curve referred to AB, AC as axes, and let

$$BAC = \theta, \quad AB = 2a, \quad AC = 2b.$$

Let x, y be the co-ordinates of E; the co-ordinates of F are a, b, and the equation of the normal at E is

$$\eta - y = \frac{y \cos\theta - x}{x \cos\theta - y}(\xi - x).$$

And if this pass through F, the co-ordinates of which are a, b,

$$(b-y)(x\cos\theta - y) = (a-x)(y\cos\theta - x),$$

or
$$x^2 - (a + b\cos\theta)x = y^2 - (a\cos\theta + b)y \quad . \quad . \quad . \quad (2)$$

The equations (1) and (2) determine all the points of the hyperbola, the tangents at which can be lines of flotation.

Also (2) is the equation to a rectangular hyperbola, referred to conjugate diameters parallel to AB, AC; the points of intersection of the two hyperbolas are therefore the positions of E.

To find x, we have

$$x^4 - (a + b\cos\theta)\, x^3 + (a\cos\theta + b)c^2 x - c^4 = 0,$$

an equation which has only one negative root, and one or three positive roots, and there may be therefore three positions of equilibrium or only one.

If the densities of the liquid and the prism be ϱ and σ, we have, since the area PAQ

$$= \tfrac{1}{2} AP \cdot AQ \sin\theta = 2xy \sin\theta = 2c^2 \sin\theta,$$

$$2\varrho c^2 \sin\theta = 2\sigma ab \sin\theta,$$

or
$$\varrho c^2 = \sigma ab,$$

from which c is determined.

Suppose the prism to be isosceles, then putting $a=b$, the equation for x becomes

$$x^4-c^4-a(1+\cos\theta)(x^3-c^2x)=0\,;$$

from which we obtain $x=c$, which gives $y=c$, and makes BC horizontal, an obvious position of equilibrium, and also

$$x=\frac{a}{2}(1+\cos\theta)\pm\left\{\frac{a^2}{4}(1+\cos\theta)^2-c^2\right\}^{\frac{1}{2}}=a\cos^2\tfrac{1}{2}\theta\pm(a^2\cos^4\tfrac{1}{2}\theta-c^2)^{\frac{1}{2}}\,;$$

the isosceles prism will therefore have only one position of equilibrium, unless

$$a\cos^2\tfrac{1}{2}\theta>c\,;$$

and since $\varrho c^2=\sigma a^2$, this is equivalent to

$$\cos^2\tfrac{1}{2}\theta>\sqrt{(\sigma/\varrho)}.$$

46. If a solid float under constraint, the conditions of equilibrium depend on the nature of the constraining circumstances, but in any case the resultant of the constraining forces must act in a vertical direction, since the other forces, the weight of the body, and the fluid pressure, are vertical.

If for instance one point of a solid be fixed, the condition of equilibrium is that the weight of the body and the weight of the fluid displaced should have equal moments about the fixed point; this condition being satisfied, the solid will be at rest, and the stress on the fixed point will be the difference of the two weights.

As an additional illustration, consider the case of a solid floating in water and supported by a string fastened to a point above the surface; in the position of equilibrium the string will be vertical, and the tension of the string, together with the resultant fluid pressure, which is equal to the weight of the displaced fluid, will counterbalance the weight of the body; the tension is therefore equal to the difference of the weights, and the weights are inversely in the ratio of the distances of their lines of action from the line of the string, these three lines being in the same vertical plane.

47. For subsequent investigations, the following geometrical propositions will be found important.

If a solid be cut by a plane, and this plane be made to turn through a very small angle about a straight line in itself, the volume cut off will remain the same, provided the straight line pass through the centroid of the area of the plane section.

To prove this, consider a right cylinder of any kind cut by a plane making with its base an angle θ.

Let $\bar{z}$ be the distance from the base of the centroid of the section

A, δA an element of the area of the section and V the volume between the planes. Then

$$\bar{z}=\frac{\Sigma(\delta A \,.\, PN)}{A};$$

$$\therefore A \cos \theta \bar{z}=\Sigma(\delta A \cos \theta \,.\, PN)=V,$$

or

$$V=\bar{z}(\text{area of base}).$$

Now the centroid of the area A is also the centroid of all sections made by planes passing through it, as may be seen by projecting the sections on the base of the cylinder; it follows, therefore, that $\bar{z}$ being the same for all such sections, the volumes cut off are the same.

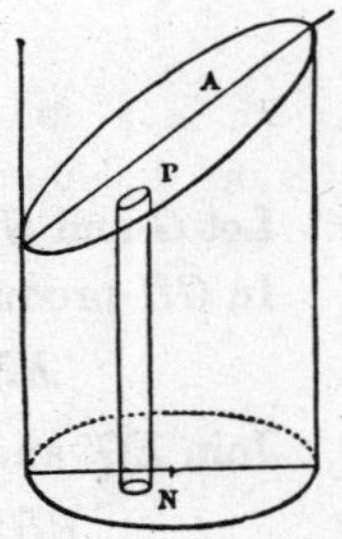

In the case of any solid, if the cutting plane be turned through a very small angle about the centroid of its section, the surface near the curves of section may be considered, without sensible error, cylindrical, and the above proposition is therefore established.

In other words, the difference between the volume lost and the volume gained by the change in the position of the cutting plane will be indefinitely small compared with either.

48. *Definitions.* If a body float in a homogeneous liquid, the plane in which the body is intersected by the surface of the liquid is the **plane of flotation.**

The point H, the centre of mass of the liquid displaced, is the **centre of buoyancy.**

If the body move so that the volume of liquid displaced remains unchanged, the envelope of the planes of flotation is the **surface of flotation,** and the locus of H is the **surface of buoyancy.**

Curves of flotation and **curves of buoyancy** are the principal normal sections at corresponding points on a surface of flotation and a surface of buoyancy.

49. *If a plane move so as to cut from a solid a constant volume, and if H be the centroid of the volume cut off, the tangent plane at H to the surface which is the locus of H is parallel to the cutting plane.*

In other words, the tangent planes at any point of the surface of flotation, and at the corresponding point of the surface of buoyancy, are parallel to one another.

Turn the plane ACB, the cutting plane, through a small angle into the position aCb, the volumes of the wedges ACa, BCb being equal.

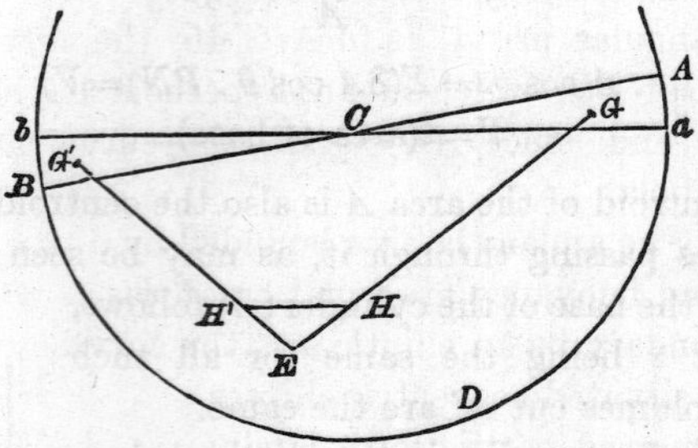

Let G and G' be the centroids of these wedges.

In GH produced take a point E such that

$$EH : HG :: \text{Volume } ACa : \text{Volume } aDB.$$

Join EG' and take H' such that

$$EH' : H'G' :: \text{Volume } BCb : \text{ Volume } aDB;$$

then H' is the centroid of aDb;

but $$EH : HG :: EH' : H'G',$$

and HH' is therefore parallel to GG'.

Hence it follows that ultimately when the angle ACa is indefinitely diminished,

$$HH' \text{ is parallel to } ACB;$$

and HH' is a tangent at H to the locus of H.

This being true for any displacement of the plane ACB about its centroid, it follows that the tangent plane at H to the locus of H is parallel to the plane ACB.

50. *The positions of equilibrium of a body floating in a homogeneous liquid are determined by drawing normals from G, the centre of mass of the body, to the surface of buoyancy.*

For if GH be a normal to the surface of buoyancy, the tangent plane at H, being parallel to the plane of flotation, is horizontal, and GH is therefore vertical.

The two conditions of equilibrium are then satisfied, and a position of equilibrium is determined.

The problem comes to the same thing as determining the positions of equilibrium of a heavy body, bounded by the surface of buoyancy, resting on a horizontal plane.

51. *Particular cases of curves of buoyancy.*

For a triangular prism, as in Art. 45, the curve of flotation is the envelope of PQ, which is an hyperbola having AB, AC for asymptotes; and, since $AH=\frac{2}{3}AE$, the curve of buoyancy is a similar hyperbola.

If the body be a plane lamina bounded by a parabola, the curves of flotation and buoyancy are equal parabolas.

If the boundary be an elliptic arc, the curves are arcs of similar and similarly situated concentric ellipses.

If the immersed portion of a lamina (or prism) be a rectangle, the curve of flotation is clearly a single point; and the curve of buoyancy is a parabola.

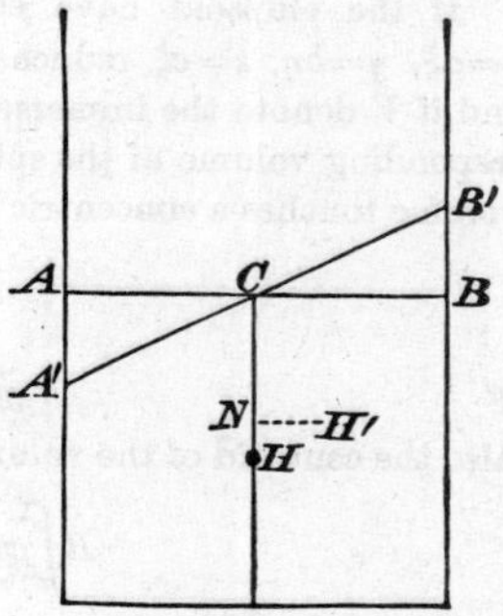

To prove the last statement, let H, H' be positions of the centroid corresponding to the positions ACB, $A'CB'$ of the line of flotation.

Then, if $AC=CB=a$, $BB'=\beta$, $CH=c$, and S=the area cut off,

$$Sy=S\,.\,H'N=\tfrac{1}{2}a\beta\,.\,\frac{2a}{3}-\tfrac{1}{2}a\beta\left(-\frac{2a}{3}\right)=\tfrac{2}{3}a^2\beta,$$

$$Sx=S\,.\,HN=\tfrac{1}{2}a\beta\left(c+\frac{\beta}{3}\right)-\tfrac{1}{2}a\beta\left(c-\frac{\beta}{3}\right)=\tfrac{1}{3}a\beta^2,$$

and $$\therefore\ Sy^2=\tfrac{4}{3}a^3x.$$

In the case of Ex. (2), Art. 45, $S=2a^2$, and the curve of buoyancy is the parabola, $3y^2=2ax$.

The radius of curvature at the vertex, H, of this parabola is $\frac{1}{3}a$, which is less than HG.

Hence it will be seen that three normals can be drawn to the curve of buoyancy, giving the three positions of equilibrium.

52. In the case of a right circular cone floating with its vertex beneath the surface, the surfaces of flotation and buoyancy are hyperboloids of revolution.

If V is the vertex of the cone, ACB the major axis of a section, and VK the perpendicular upon AB, the volume VAB is equal to

$$\tfrac{1}{3}VK\,.\,\tfrac{1}{2}\pi AB\,.\,\{AV\,.\,BV\sin^2\alpha\}^{\frac{1}{2}}.$$

But $$VK\,.\,AB=VA\,.\,VB\sin 2\alpha,$$

each expression being double the area VAB; therefore, the volume being constant, it follows that the area VAB is constant.

The locus of C', the centroid of the plane section, is therefore a hyperboloid of revolution, and, VH being three-fourths of VC, the surface of buoyancy is a similar hyperboloid.

53. Surfaces of buoyancy and flotation for an ellipsoid.

If the ellipsoid have equation $x^2/a^2+y^2/b^2+z^2/c^2=1$, the substitutions $x=a\xi$, $y=b\eta$, $z=c\zeta$ reduce the problem to that of the sphere $\xi^2+\eta^2+\zeta^2=1$; and if V denote the immersed volume of the ellipsoid, V/abc denotes the corresponding volume of the sphere. It is clear that the plane which cuts off this volume touches a concentric sphere of radius r, such that

$$\int_r^1 \pi(1-x^2)dx=V/abc,$$

or

$$\tfrac{1}{3}\pi(1-r)^2(2+r)=V/abc.$$

Also the centroid of the volume cut off lies on a sphere of radius R, where

$$R\int_r^1 \pi(1-x^2)dx=\int_r^1 \pi x(1-x^2)dx$$

or

$$R=\tfrac{3}{4}(1+r)^2/(2+r).$$

Returning to the original problem, we see that the surface of flotation is a similar ellipsoid of semiaxes ra, rb, rc, where

$$(1-r)^2(2+r)=3V/\pi abc \quad . \quad . \quad . \quad . \quad (1)$$

and the surface of buoyancy is another similar ellipsoid of semiaxes Ra, Rb, Rc, where

$$R=\tfrac{3}{4}(1+r)^2/(2+r) \quad . \quad . \quad . \quad . \quad (2)$$

Similar results hold good for a hyperboloid of two sheets.

54. Elliptic Paraboloid.

This case can be deduced from the results for an ellipsoid by making a, b, c tend to infinity in such a way that $a^2/c\to\alpha$ and $b^2/c\to\beta$, where α, β are the semi latera recta of the principal sections of the paraboloid. If, as before, V denotes the finite volume immersed, then V/abc tends to zero, so that r and also R both tend to unity. Hence the surfaces of flotation and buoyancy are equal paraboloids. Also the distances between their vertices and the vertex of the given paraboloid are the limiting values of $c(1-r)$ and $c(1-R)$.

But from Art. 53 (1), we see that

$$c^2(1-r)^2=\frac{3Vc}{(2+r)\pi ab}\to\frac{V}{\pi\sqrt{\alpha\beta}};$$

so that the intercept on the axis between the given paraboloid and the surface of flotation is γ, where

$$\gamma^2 = V/\pi\sqrt{\alpha\beta}.$$

Similarly, from Art. 53 (2),

$$c(1-R) = \frac{c(1-r)(5+3r)}{4(2+r)} \rightarrow \tfrac{8}{3}\gamma,$$

thus determining the corresponding intercept for the surface of buoyancy.

55. Cylinder of any section.

The surface of flotation is a point on the line of centroids Oz, given by $Ac=V$, where A is the cross-section and V the volume immersed.

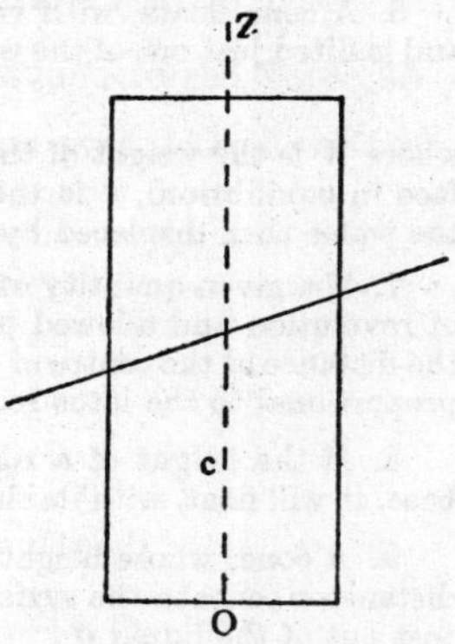

Let $z=lx+my+c$ be the equation of the cutting plane, the origin being in the base.

The co-ordinates $(\bar{x}, \bar{y}, \bar{z})$ of the centre of buoyancy are given by

$$V\bar{x} = \iint xzdxdy \text{ integrated over the base}$$
$$= \iint x(c+lx+my)dxdy$$
$$= al+hm.$$

Similarly

$$V\bar{y} = \iint yzdxdy$$
$$= hl+bm;$$

and

$$V\bar{z} = \tfrac{1}{2}\iint z^2dxdy$$
$$= \tfrac{1}{2}(al^2+2hlm+bm^2)+\tfrac{1}{2}c^2A;$$

where $a=\iint x^2dxdy,\quad h=\iint xydxdy,\quad b=\iint y^2dxdy.$

If we use the principal axes of the section as axes of x and y, we have $h=0$, and

$$V\bar{x}=al,\quad V\bar{y}=bm,\quad V(\bar{z}-\tfrac{1}{2}c)=\tfrac{1}{2}(al^2+bm^2).$$

Therefore the equation of the surface of buoyancy is

$$\frac{x^2}{a}+\frac{y^2}{b}=\frac{2z-c}{V}.$$

EXAMPLES

1. A solid formed of two co-axial right cones, of the same vertical angle, connected at the vertices, is placed with one end in contact with the horizontal base of a vessel: water is then poured into the vessel; show that if the altitude of the upper cone be treble that of the lower, and the common density of the spindle four-sevenths that of the water, it will be upon the point of rising when the water reaches to the level of its upper end.

2. A cone, of given weight and volume, floats with its vertex downwards; prove that the surface of the cone in contact with the liquid is least when its vertical angle is $2\tan^{-1}1/\sqrt{2}$.

3. A hollow hemispherical shell has a heavy particle fixed to its rim, and floats in water with the particle just above the surface, and with the plane of

the rim inclined at an angle of 45° to the surface ; show that the weight of the hemisphere :. the weight of the water which it would contain

$$:: 4\sqrt{2}-5 : 6\sqrt{2}.$$

4. A solid cone is divided into two parts by a plane through its axis, and the parts are connected by a hinge at the vertex ; the system being placed in water with its axis vertical and vertex downwards, show that, if it float without separation of the parts, the length of the axis immersed is greater than $h \sin^2 a$, h being the height of the cone, and $2a$ its vertical angle.

5. A cylinder floats in a liquid with its axis inclined at an angle $\tan^{-1}2/5$ to the vertical, and its upper end just above the surface ; prove that the radius is 4/7 of the height of the cylinder.

6. A cone floats, with vertex downwards, in a cylindrical basin of water, and is lifted just out of the water (without tilting) ; show that the work done is

$$W(\tfrac{3}{4}l-\tfrac{1}{3}l'),$$

where W is the weight of the cone, l is the depth of the vertex below the surface in equilibrium, l' is the length of the cylinder which would be filled by the water then displaced by the cone.

7. If a given quantity of homogeneous matter be formed into a paraboloid of revolution and allowed to float with the vertex downwards, the square of the distance of the centre of gravity from the plane of flotation will be inversely proportional to the latus rectum.

8. If the height of a right circular cone be equal to the diameter of the base, it will float, with its slant side horizontal, in any liquid of greater density.

9. A cone, whose height is h and vertical angle $2a$, has its vertex fixed at distance c beneath the surface of a liquid ; show that it will rest with its base just out of the liquid if

$$\sigma c^4 \cos^3 a \cos \theta = \varrho h^4 [\cos(\theta-a)\cos(\theta+a)]^{\frac{3}{2}},$$

where σ and ϱ are the densities of the liquid and cone, and θ is given by the equation $c \cos a = h \cos(\theta+a)$.

10. A right circular cylinder, whose axis is vertical, contains a quantity of liquid, the density of which varies as the depth, and a right cone whose axis is coincident with that of the cylinder and which is of equal base, is allowed to sink slowly into the liquid with its vertex downwards. If the cone be in equilibrium when just immersed, prove that the density of the cone is equal to the initial density of the liquid at a depth equal to $\frac{1}{12}$th the length of the axis of the cone.

11. A solid cone, of height h, vertical angle $2a$, and density ϱ, is movable about its vertex, and its vertex is fixed at a depth c below the surface of a liquid, the density of which, at a depth z, is μz. The cone is in equilibrium with its axis inclined at an angle θ to the vertical, and its base above the surface ; prove that

$$\mu c^5 \cos^3 a \cos \theta = 5\varrho h^4 \{\cos(\theta+a)\cos(\theta-a)\}^{\frac{3}{2}}.$$

12. A hollow paraboloidal vessel floats in water with a heavy sphere lying in it. There being an opening at the vertex, the water occupies the whole of the space between the vessel and the sphere. If the resultant pressure on the sphere be equal to half the weight of the water which would fill it, show that the depth of the centre of the sphere below the surface of the water is $4a^2/3c$, where $4a$ is the latus rectum of the paraboloid, and c the distance of the plane of contact from the vertex.

13. A right-angled triangular prism floats in a fluid of which the density varies as the depth with the right angle immersed and the edges horizontal; show that the curve of buoyancy is of the form

$$r^3 \sin^2\theta \cos^2\theta = c^3.$$

14. A life-belt in the form of an anchor-ring generated by a circle of radius a floats in water with its equatorial plane horizontal; show that z, the depth immersed, is given by the equations

$$z = a(1-\cos\beta),$$

$$2\pi s = (2\beta - \sin 2\beta);$$

where s is the specific gravity of the material of the belt.

15. An indefinitely small piece of ice, the shape of which may be taken to be that of a right circular cylinder, is floating in water with its axis vertical. The part immersed receives deposits of ice in such a manner as to continue cylindrical, the radius and axis receiving equal increments in equal times. Find the ultimate shape of the part not immersed.

If the specific gravity of ice be ·96, prove that the surface is formed by the revolution of the curve

$$y^2(9x-y)^{25} = a^{27}.$$

16. A solid bounded by the planes $x=\pm a$, $y=\pm b$, $z=0$, and $z=c$ floats in water with the base $z=0$ wholly immersed. Show that for displacements such that the volume V immersed remains constant and the base is entirely under water and the opposite face entirely out of the water, the equation of the surface of buoyancy is

$$\frac{x^2}{a^2}+\frac{y^2}{b^2}=\frac{8abz}{3V}-\frac{1}{3}.$$

17. A cylindrical vessel with its cross-section of any shape floats with a length $2c$ of its axis immersed when the axis is vertical. Prove that the equation of the surface of buoyancy is $x^2/a^2+y^2/b^2=z/c$; where the origin is taken at the middle point of the portion of the axis immersed for the upright position, the axis of z is vertically upwards, and the axes of x, y parallel to the principal axes of moments of inertia of the plane of flotation for the upright position through its centre of gravity, and b, a are the radii of gyration for those axes of the plane of flotation.

CHAPTER V

THE STABILITY OF THE EQUILIBRIUM OF FLOATING BODIES

56. If a floating body be slightly displaced it will in general either tend to return to its original position or will recede farther from that position ; in the former case the equilibrium is said to be *stable*, and in the latter *unstable*, for that particular direction of displacement.

Consider first a small vertical displacement : it is clear that, if the body be floating partially immersed in homogeneous fluid, or if it be immersed, either wholly or partially, in a heterogeneous fluid of which the density increases with the depth, a depression will increase the weight of the fluid displaced, and on the contrary an elevation will diminish it ; in either case the tendency of the fluid pressure is to restore the body to its position of rest, and the equilibrium is stable with regard to vertical displacements. This, it will be observed, is only shown to be true of rigid bodies ; if the increased pressure, caused by depression, have the effect of compressing any portion of the floating body, the equilibrium is not necessarily stable, and in fact it may be unstable.

An arbitrary displacement will in general involve both vertical and angular changes in the position of the body ; if however the displacement be small, as we have supposed to be the case, the effects of the two changes of position can be treated independently ; and we proceed to consider the effect of a small angular displacement, on the supposition that the weight of fluid displaced remains unchanged, and consequently that the fluid pressure has no tendency to raise or depress the centre of mass of the body.

57. *A solid, floating at rest in a homogeneous liquid, is made to turn through a small angle in a given vertical plane ; to determine whether the fluid pressure will tend to restore it to its original position or not.*

Suppose that the body is turned through a small angle θ about an axis Oy in the plane of flotation AOB; Oy being at right angles to the plane of the paper, Ox in the plane of flotation and Oz vertical in the original position; and as the body is turned let the axes be carried with it.

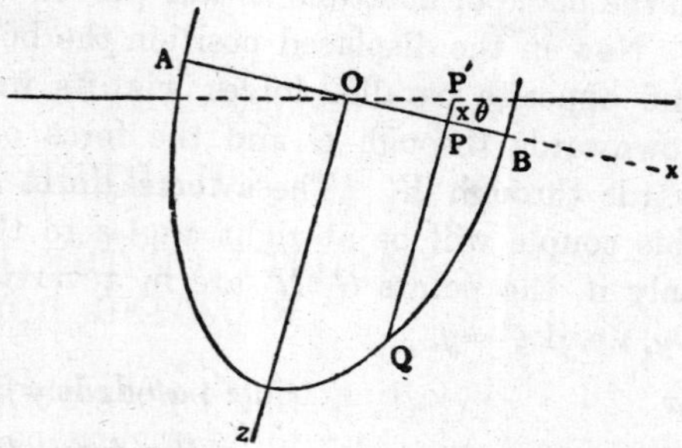

If $dxdy$ denotes an element of area on the plane of flotation AOB, the volume of an elementary column PQ is $zdxdy$, where z denotes the length PQ. In the displaced position the length of the corresponding column $P'Q$ is $z+x\theta$ and its volume is $(z+x\theta)dxdy$. Hence the volume V of liquid displaced will be the same in both cases if

$$\iint(z+x\theta)dxdy=V=\iint zdxdy,$$

where the integrations are over the section of the body made by the plane of flotation in the original position.

This reduces to $\iint xdxdy=0$, which means that the centre of gravity of the surface section must lie on Oy, as was proved in Art. 47.

Assume that this condition is satisfied. In the original position the centre of gravity G and centre of buoyancy H are in the same vertical, and we may denote the co-ordinates of the latter by $(\bar{x}, \bar{y}, \bar{z})$ and note that G will have the same $(\bar{x}, \bar{y})$. In the displaced position there is a new centre of buoyancy H' whose co-ordinates referred to the original axes are $(\bar{x}', \bar{y}', \bar{z}')$.

$$\text{Now } V\bar{x}=\iint xzdxdy, \quad V\bar{y}=\iint yzdxdy, \quad V\bar{z}=\iint \tfrac{1}{2}z^2dxdy.$$

These integrals being written down by taking the elementary column PQ of volume $zdxdy$ with its centre of gravity at the middle point of its length.

In the displaced position the corresponding elementary column is $P'Q$ of length $z+x\theta$; its centre of gravity is at a distance $\frac{1}{2}(z+x\theta)$ from P', and therefore at a distance $\frac{1}{2}(z-x\theta)$ from P, so that we have

$$V\bar{x}'=\iint x(z+x\theta)dxdy, \quad V\bar{y}'=\iint y(z+x\theta)dxdy,$$
$$V\bar{z}'=\iint \tfrac{1}{2}(z-x\theta)(z+x\theta)dxdy.$$

We observe that, to the first power of the small angle θ, we have $\bar{z}'=\bar{z}$, so that the tangent plane to the surface of buoyancy is parallel to the plane of flotation, as was proved in Art. 49.

Now in the displaced position the body is subject to two equal and opposite parallel forces, viz. its weight W or $g\rho V$ vertically downwards through G and the force of buoyancy vertically upwards through H'. These forces form a couple and the plane of this couple will be at right angles to the axis of rotation if, and only if, the points G, H' are in a vertical plane perpendicular to Oy, *i.e.* if $\bar{y}'=\bar{y}$,

or
$$\iint y(z+x\theta)dxdy=\iint yzdxdy.$$

This reduces to
$$\iint xydxdy=0,$$

which means that the axis of rotation Oy must be a principal axis of inertia of the section of the body made by the plane of flotation.

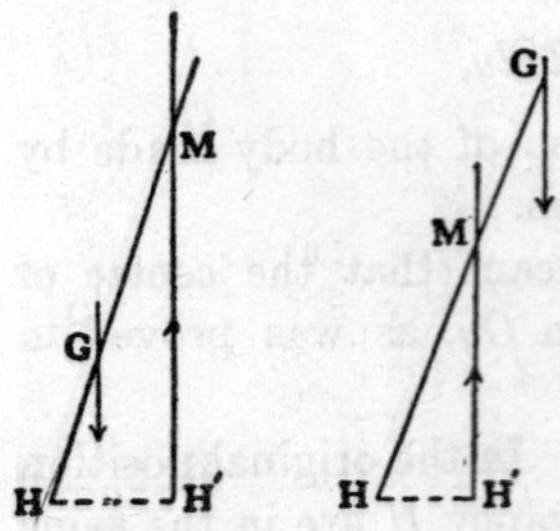

When this condition is satisfied the vertical through H' intersects the line HG in a point M called the **metacentre.** The couple acting on the body is $W \cdot GM\theta$, and it tends to restore the body to its former position or to increase the displacement according as M is above or below G.

Also, we have
$$HM \cdot \theta=HH'=\bar{x}'-\bar{x}$$
$$=\frac{\theta\iint x^2dxdy}{V}.$$

Therefore $HM=Ak^2/V$, where Ak^2 denotes the moment of inertia of the section of the body made by the plane of flotation about the axis of rotation.

The couple tending to restore the body is therefore
$$g\rho\theta V(HM-HG)=g\rho\theta(Ak^2-V \cdot HG).$$

58. Since there are two principal axes through the centre of gravity of the surface section of the body with corresponding moments of inertia I_1 and I_2, it follows that a displacement about either of these axes would set up a couple in the plane of the displacement tending to restore equilibrium if $GH < I_1/V$ and also $< I_2/V$. Hence these conditions are necessary for stability of equilibrium.

59. Work done in producing a displacement. When the body has been displaced through a small angle θ about either principal axis through the centre of gravity of the surface section, the couple acting on the body is

$$g\rho(Ak^2-V\,.\,HG)\theta.$$

Consequently the work that would have to be done by external agency in order to increase θ by a small amount $d\theta$ is

$$g\rho(Ak^2-V\,.\,HG)\theta d\theta,$$

and, by integration, it follows that the work done in producing the angular displacement θ is

$$\tfrac{1}{2}g\rho(Ak^2-V\,.\,HG)\theta^2.$$

60. Sufficiency of the conditions for stability. A small rotation about any axis in the plane of flotation through the centre of gravity of the water-section may be regarded as compounded of rotations θ_1, θ_2 about the principal axes of the section. Each of these separately sets up a restoring couple, and the total work that would have to be done by external agency, or the gain in potential energy, in producing the displacement is *

$$\tfrac{1}{2}g\rho(I_1-V\,.\,HG)\theta_1^2+\tfrac{1}{2}g\rho(I_2-V\,.\,HG)\theta_2^2.$$

Whence it follows that the conditions $HG < I_1/V$ and also $< I_2/V$ are sufficient to ensure stability for displacements which do not alter the volume of liquid displaced.

61. The question of stability may also be treated somewhat differently.

Defining a *metacentre* as the point of intersection with the line HG of the vertical line through the new centre of buoyancy after a slight displacement, we are led to the following theorem :

A **metacentre** *is a centre of curvature of the surface of buoyancy at the point in the same vertical line with G.*

This is at once obvious from the fact that the point M is the point of intersection of consecutive normals to the surface.

Hence it appears that for any displacement, consistent with the conditions for the existence of a metacentre, the direction of the fluid pressure is always a vertical tangent to the evolute of the curve of buoyancy.

* That the expression for the work done in a displacement of this kind does not contain a term $\theta_1\theta_2$ may be *proved* as in Art. 66 following.

62. A most important case naturally presents itself; that is, the question of the stability of equilibrium of a ship when displaced by rolling.

In general it is impossible for a ship to roll without tossing, because the two ends of the ship are unsymmetrical; but in the case of a very long vessel, such as an Atlantic "liner," it may be

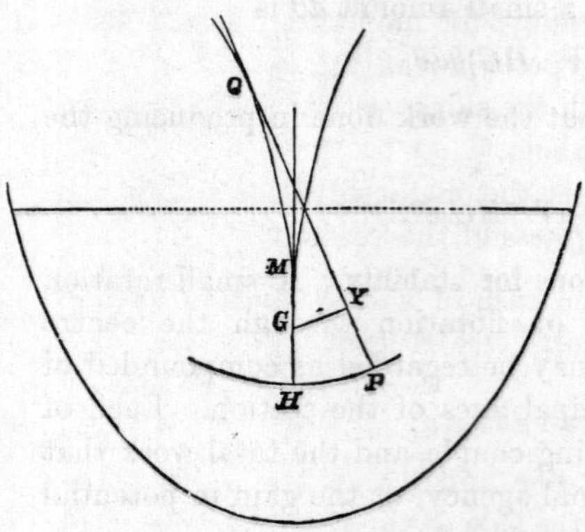

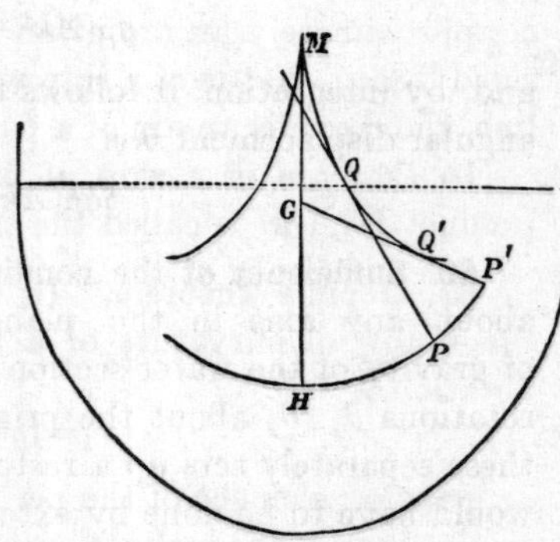

assumed that the ship can be divided symmetrically by a plane perpendicular to its length, and in this case the ship has two vertical planes of symmetry, and consequently the vertical line HG passes through the centroid of the plane of flotation.

The line HG also divides the curves of buoyancy symmetrically, and the point H is a point of maximum or minimum curvature. In the first of these two cases the cusp of the evolute is pointed downwards; in the second case it is pointed upwards.

The figures at once show the effects of displacement.

In the first case the righting moment, which is the statical measure of stability for a given angle of displacement, is proportional to GY the perpendicular from G on the tangent PQ, and increases with an increase in the angle of displacement.

In the second case the righting moment increases to a maximum value, and then diminishes, vanishing for the position given by the tangent $GQ'P'$.

This is a position of equilibrium, but it is of unstable equilibrium, in accordance with the general mechanical law that positions of stable and unstable equilibrium occur alternately.

If the equation to the curve of buoyancy be obtained in the form $p=f(\phi)$, G being the origin,

$$GY=dp/d\phi,$$

and the righting moment is

$$Wdp/d\phi,$$

if W be the weight of the ship.

In general the curve of buoyancy, for moderate displacements, is approximately an arc of an hyperbola; in the case of a "wall-sided" ship, that is of a ship with the sides vertical near the water-line, the curve is an arc of a parabola.

In the case of a ship, if M is the metacentre for rolling, the product $W \,.\, GM$ is called the *stiffness* of the vessel.

63. Dupin's Theorem. In the case of a ship floating upright, the radius of curvature of a transverse section of the surface of flotation is

$$r_1=\int y^2 \tan a\,ds/A,$$

ds being an element of the perimeter, and A the area, of the water-section, and a the inclination of the side of the ship to the vertical; the axes of x and y being the longitudinal and transverse axes of the section of the vessel by the plane of flotation through its centroid C.

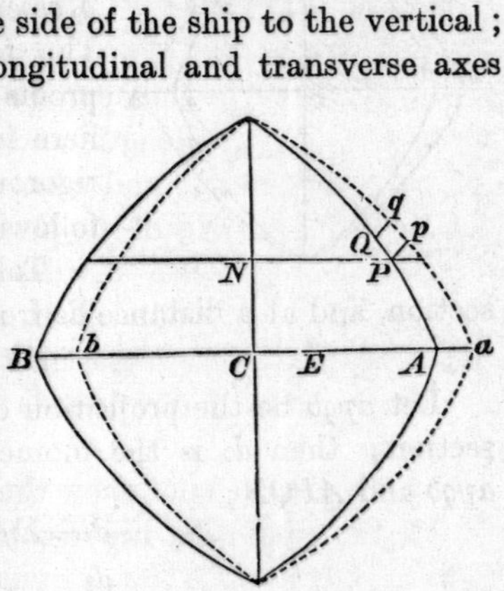

To prove this let C, C' be neighbouring points on the transverse section of the surface of flotation, the tangent plane at C' making a small angle θ with the water-section $APQB$, and let $apqb$ be the projection on the water-section of the section of the ship made by this tangent plane, so that E, the projection of C', is the centroid of the area $apqb$. Let PQ, pq be corresponding elements, and $PQ=ds$, then

$$\text{area } PQpq=y\theta \tan a\,ds;$$

$$\therefore CE \,.\, (A)=\int y^2\theta \tan a\,ds,$$

and, since $CC'=r_1\theta$, and $CE=CC'$ ultimately, it follows that

$$r_1A=\int y^2 \tan a\,ds,$$

an expression first given by C. Dupin, in a memoir presented to the Académie des Sciences in 1814. A corresponding expression obviously exists for the radius of curvature (R_1) of the longitudinal section.

64. Leclert's Theorem. Calling r and R the metacentric heights for transverse and longitudinal displacements, that is, the radii of curvature of transverse and longitudinal sections of the surface of buoyancy; we know that

$$r=\frac{i}{V} \text{ and } R=\frac{I}{V},$$

where i and I are the principal moments of inertia of the water-section. E. Leclert has established the following relations between these quantities:

$$r_1=\frac{di}{dV}=r+V\frac{dr}{dV};\ R_1=\frac{dI}{dV}=R+V\frac{dR}{dV}.$$

A translation of Leclert's paper is given by Mr Merrifield in the *Proceedings*, for 1870, *of the Institution of Naval Architects*, and in the *Messenger of Mathematics*, March 1872. The following is the first of the two proofs which are given; it is retained here for its historic interest, but a more rigorous treatment is given in Art. 67 following.

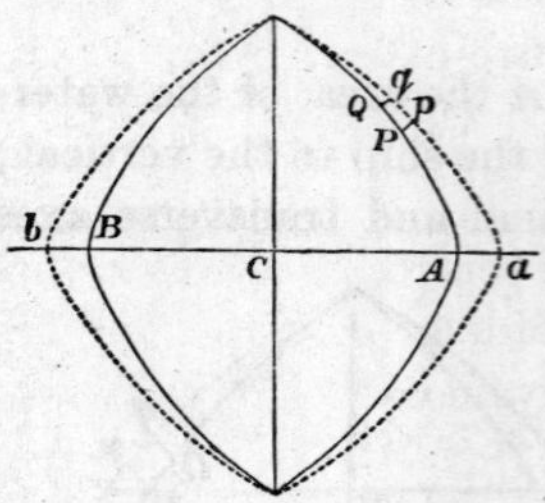

Taking a section parallel to the water-section, and at a distance dz from it,

$$dV=Adz.$$

Let $apqb$ be the projection of this new section upon the water-section; then di is the moment of inertia of the area between $apqb$ and $APQB$;

$$\therefore di=\Sigma y^2 dz \,.\, \tan ads,$$

and

$$\frac{di}{dz}=\int y^2 \tan ads.$$

Hence

$$r_1=\frac{1}{A}\frac{di}{dz}=\frac{di}{dV};$$

$$\therefore r_1-r=\frac{di}{dV}-\frac{i}{V}=V\frac{d}{dV}\left(\frac{i}{V}\right),$$

or

$$r_1=r+V\frac{dr}{dV}.$$

65. Surface of buoyancy in general.

Let the origin be taken in the vertical through the centroid of the original water-line section. Then if $z=c$ be the original section, the plane in the slightly displaced position will be

$$z=c+lx+my$$

where l, m are small.

If (x_0, y_0, z_0) and (x, y, z) denote the co-ordinates of the centre of buoyancy in the two positions

$$V(x-x_0)=\iint(z-c)x\,dx\,dy=al+hm,$$
$$V(y-y_0)=\iint(z-c)y\,dx\,dy=hl+bm,$$
$$V(z-z_0)=\iint\tfrac{1}{2}(z^2-c^2)\,dx\,dy=\tfrac{1}{2}(al^2+2hlm+bm^2),$$

where $a=\iint x^2dx\,dy,\ h=\iint xy\,dx\,dy,\ b=\iint y^2dx\,dy.$

Hence $\quad 2(z-z_0)=l(x-x_0)+m(y-y_0)$

or $2(z-z_0)=\dfrac{V}{ab-h^2}\{b(x-x_0)^2-2h(x-x_0)(y-y_0)+a(y-y_0)^2\}$

is the approximate form of surface of buoyancy. If the original axes of x and y are principal axes of the plane section, then $h=0$, and if the origin be now moved to the centre of buoyancy in the first position, the surface becomes

$$2z=Vx^2/a+Vy^2/b.$$

If we now define the **metacentres** as the centres of curvature of the principal normal sections of the surface of buoyancy, the heights of the metacentres above the centre of buoyancy are the principal radii of curvature a/V or b/V.

66. Condition for stability.

The tangent plane to the surface of buoyancy at a point (x, y, z) is given by

$$\zeta-z=\frac{Vx}{a}(\xi-x)+\frac{Vy}{b}(\eta-y).$$

And the perpendicular distance of the centre of gravity $(0, 0, \bar{z})$ of the solid from this plane is

$$\left\{\bar{z}-z+\frac{Vx^2}{a}+\frac{Vy^2}{b}\right\}\left\{1+\frac{V^2x^2}{a^2}+\frac{V^2y^2}{b^2}\right\}^{-\frac{1}{2}}$$

$$=\left\{\bar{z}+\frac{Vx^2}{2a}+\frac{Vy^2}{2b}\right\}\left\{1-\frac{V^2x^2}{2a^2}-\frac{V^2y^2}{2b^2}\right\}$$

$$=\bar{z}+\frac{V^2x^2}{2a^2}\left(\frac{a}{V}-\bar{z}\right)+\frac{V^2y^2}{2b^2}\left(\frac{b}{V}-\bar{z}\right).$$

Now by Art. 50 the positions of equilibrium correspond to those of a heavy body bounded by the surface of buoyancy resting on a horizontal plane, so that for stability the height of the centre of gravity above the plane must be a minimum. This requires that $\bar{z}$ should be less than $\frac{a}{V}$ and $\frac{b}{V}$, or the centre of gravity must be below both metacentres.

67. Surface of Flotation. Leclert's Theorem.

Suppose that the volume immersed is increased by a small amount δV by depressing the solid from the second position of Art. 65.

If ξ, η, ζ are the co-ordinates of the centre of gravity of the thin slice, of volume δV, since $al+hm$=difference of x-moments of volume displaced, therefore by Art. 65,

$$l\delta a+m\delta h=\xi\delta V.$$

Similarly $$\eta\delta V=l\delta h+m\delta b;$$

and $$\zeta\delta V=\tfrac{1}{2}(l^2\delta a+2lm\delta h+m^2\delta b).$$

Also as the thickness of the slice is diminished the point (ξ, η, ζ) tends to coincide with the corresponding point on the surface of flotation, *i.e.* the centroid of the water-line area.

Hence on the surface of flotation we have

$$x'.dV=lda+mdh$$
$$y'.dV=ldh+mdb$$
$$z'.dV=\tfrac{1}{2}(l^2da+2lmdh+m^2db),$$

and its equation is

$$2z'=\frac{dV}{dadb-(dh)^2}\{x'^2db-2x'y'dh+y'^2da\}.$$

In the special case in which $dh=0$, this becomes

$$2z'=x'^2\frac{dV}{da}+y'^2\frac{dV}{db},$$

and the radii of curvature of the surface of flotation are $\frac{da}{dV}$ and $\frac{db}{dV}$ as in Art. 64.

We observe that the principal axes of two parallel sections of the solid are not necessarily parallel, so that $h=0$ does not imply that $dh/dV=0$. The results of Art. 64 are thus seen to be true only in the cases there implied in which there are vertical planes of symmetry which contain all principal axes of horizontal sections.*

68. We now append some examples of the determination of the metacentre.

EXAMPLE 1. *A solid cylinder of radius* a *and length* h *floating with its axis vertical.*

In this case the plane of flotation is a circular area, and

$$Ak^2 = \tfrac{1}{4}\pi a^4;$$

therefore, if h' be the length of the axis immersed,

$$\pi a^2 h' \,.\, HM = \tfrac{1}{4}\pi a^4, \text{ or } HM = a^2/4h',$$

and the equilibrium is stable if

$$\frac{a^2}{4h'} > \frac{h}{2} - \frac{h'}{2}.$$

EXAMPLE 2. *A cylinder floating with its axis horizontal and in the surface is displaced in the vertical plane through the axis.*

The plane of flotation is a rectangle, and

$$Ak^2 = \tfrac{1}{6}ah^3,$$

h being the length of the cylinder, and a its radius;

$$\therefore HM = \frac{1}{3}\frac{h^2}{\pi a};$$

and the equilibrium is stable, if

$$\frac{1}{3}\frac{h^2}{\pi a} > \frac{4a}{3\pi},$$

or

$$h > 2a.$$

EXAMPLE 3. *A solid cone floating with its axis vertical and vertex downwards.*

Let h be the length of the axis,

z the portion of the axis immersed,

2α the vertical angle of the cone.

Then
$$Ak^2 = \tfrac{1}{4}\pi z^4 \tan^4 \alpha,$$
and
$$V = \tfrac{1}{3}\pi z^3 \tan^2 \alpha;$$
$$\therefore HM = \tfrac{3}{4}z \tan^2 \alpha;$$
also
$$HG = \tfrac{3}{4}h - \tfrac{3}{4}z,$$
and therefore the equilibrium is stable or unstable, according as
$$z \tan^2 \alpha > \text{or} < h - z,$$
or
$$z > \text{or} < h \cos^2 \alpha.$$

But if ρ, σ be the densities of the fluid and cone,

$$\left(\frac{z}{h}\right)^3 = \frac{\sigma}{\rho};$$

* This correction to Leclert's Theorem and the method of treatment of the last few Articles, as well as Arts. 76–78 below, are due to Dr Bromwich.

therefore the equilibrium is stable or unstable as

$$\frac{\sigma}{\varrho} > \text{or} < \cos^6 \alpha.$$

EXAMPLE 4. *An isosceles triangular prism floating with its base not immersed, and its edges horizontal.*

Referring to Art. 45, consider first the position of equilibrium in which the base is inclined to the horizon.

In this case, if $AQ=2y$ and $AP=2x$, and we put $a=b$ in equation (2) on page 45, x and y are given by the equations

$$x+y=2a\cos^2\tfrac{1}{2}\theta,$$
$$xy=c^2.$$

The co-ordinates of G and H referred to AB, AC as axes are respectively

$$\tfrac{2}{3}a,\ \tfrac{2}{3}a, \text{ and } \tfrac{2}{3}x,\ \tfrac{2}{3}y,$$

$$\therefore HG^2=\tfrac{4}{9}\{(a-x)^2+(a-y)^2+2(a-x)(a-y)\cos\theta\}$$
$$=\tfrac{4}{9}\{x^2+y^2+2xy\cos\theta-2a(1+\cos\theta)(x+y)+2a^2(1+\cos\theta)\},$$

from which, by means of the above equations, we obtain

$$HG=\frac{4}{3}\sin\tfrac{1}{2}\theta(a^2\cos^2\tfrac{1}{2}\theta-c^2)^{\frac{1}{2}}.$$

The area $PAQ=2c^2\sin\theta$, and if M be the metacentre, and l the length of the prism,

$$2lc^2\sin\theta\,.\,HM=\tfrac{1}{12}PQ^3\,.\,l,$$

$$\therefore HM=\frac{PQ^3}{24c^2\sin\theta}.$$

But

$$PQ^2=4(x^2+y^2-2xy\cos\theta)$$
$$=16\cos^2\tfrac{1}{2}\theta(a^2\cos^2\tfrac{1}{2}\theta-c^2);$$

$$\therefore HM=\frac{4}{3}\cos^3\tfrac{1}{2}\theta(a^2\cos^2\tfrac{1}{2}\theta-c^2)^{\frac{3}{2}}/c^2\sin^2\tfrac{1}{2}\theta,$$

and

$$HM>HG, \text{ if } c^2\sin^3\tfrac{1}{2}\theta<\cos^3\tfrac{1}{2}\theta(a^2\cos^2\tfrac{1}{2}\theta-c^2),$$

i.e. if

$$\cos^2\tfrac{1}{2}\theta>c/a.$$

Next, consider the case in which the base is horizontal, and PQ therefore parallel to BC.

The area $PAQ=2c^2\sin\theta$,

$$AP=AQ=2c, \text{ and } PQ=4c\sin\tfrac{1}{2}\theta.$$

Hence,

$$HM=\frac{4}{3}c\sin^2\tfrac{1}{2}\theta/\cos\tfrac{1}{2}\theta, \text{ and } HG=\frac{4}{3}(a-c)\cos\tfrac{1}{2}\theta,$$

and

$$HM>HG \text{ if } \cos^2\tfrac{1}{2}\theta<c/a.$$

Now in the Art. before referred to, we have shown that there are three positions of equilibrium, or one only, according as

$$\cos^2\tfrac{1}{2}\theta> \text{or} <c/a.$$

Hence it follows, that when there are three positions of equilibrium, the intermediate one, in which CB is horizontal, is a position of unstable equilibrium, while in the other two positions the equilibrium is stable.

If there be only one position in which the prism will rest, its equilibrium is stable.

It will be a useful exercise for the student to obtain these results by investigating the equation to the curve of buoyancy, and determining the position of its centre of curvature.

69. Finite displacements. If a solid body, floating in water, be turned through any given angle from its position of equilibrium, then, as before, the moment of the fluid pressure is restorative or not according as the point L at which the vertical through the new centre of buoyancy meets the line HG is above or below G, assuming these lines to intersect.

It is not to be inferred that if L is above G, the body will when set free return to its original position and oscillate through it, or even that the original position is one of stable equilibrium, according to our previous definition of stability: it is a general law of mechanics that positions of stable and unstable equilibrium occur alternately, and the body may have been displaced from its original position *through* other positions of equilibrium.

As a particular example take the following.

A solid cone, floating with its axis vertical and vertex downwards, is turned through an angle θ in a vertical plane, the volume of fluid displaced remaining the same; to determine the direction of the moment of the fluid pressure.

Let AB be the major axis of the elliptic section made by the surface plane of the fluid, C its middle point, Aa, Bb, Cc lines at right angles to AB, and let the angle $AVB=2\alpha$ and $VA=d$. Then

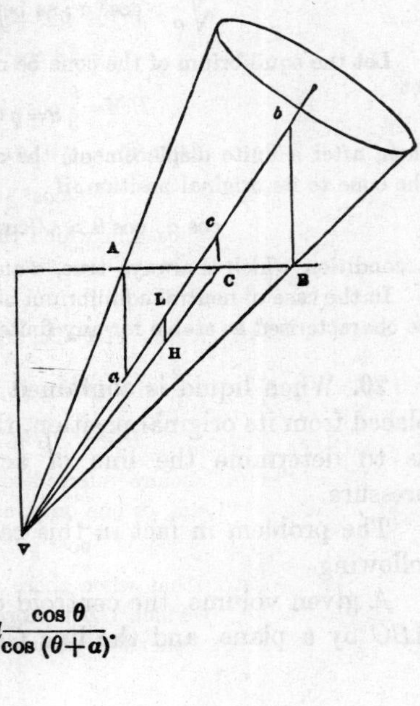

$$VAa=\theta-\alpha,$$

and $$VBb=\pi-\theta-\alpha.$$

$$Vc=\frac{1}{2}(Va+Vb)=\frac{1}{2}\cdot\left\{d\frac{\sin(\theta-\alpha)}{\sin\theta}+d\frac{\cos(\theta-\alpha)}{\cos(\theta+\alpha)}\frac{\sin(\theta+\alpha)}{\sin\theta}\right\}$$

$$=\frac{d\cos\theta}{\cos(\theta+\alpha)};$$

$$\therefore VL=\frac{3}{4}d\frac{\cos\theta}{\cos(\theta+\alpha)}.$$

The semi-minor axis of the ellipse AB is a mean proportional between the perpendiculars from A and B on the axis of the cone,

$$\therefore \text{ its area} = \pi\tfrac{1}{2}AB(VA \cdot VB \cdot \sin^2 \alpha)^{\frac{1}{2}}$$

$$= \frac{\pi}{2}d^2 \frac{\sin \alpha \sin 2\alpha}{\cos(\theta+\alpha)} \cdot \left\{\frac{\cos(\theta-\alpha)}{\cos(\theta+\alpha)}\right\}^{\frac{1}{2}};$$

therefore the volume of the fluid displaced

$$= \tfrac{1}{3}d\cos(\theta-\alpha) \cdot (\text{area of ellipse})$$

$$= \tfrac{1}{3}\pi d^3 \sin^2 \alpha \cos \alpha \left\{\frac{\cos(\theta-\alpha)}{\cos(\theta+\alpha)}\right\}^{\frac{3}{2}}.$$

Hence, if ρ, σ be the densities of the fluid and the cone, since the weight of the fluid displaced is equal to that of the cone, we have

$$\rho d^3 \sin^2 \alpha \cos \alpha \left\{\frac{\cos(\theta-\alpha)}{\cos(\theta+\alpha)}\right\}^{\frac{3}{2}} = \sigma h^3 \tan^2 \alpha,$$

or
$$\left(\frac{d}{h}\right)^3 = \frac{\sigma}{\rho}\left\{\frac{\cos(\theta+\alpha)}{\cos(\theta-\alpha)}\right\}^{\frac{3}{2}} \frac{1}{\cos^3 \alpha}.$$

And $VL > VG$ if
$$d\frac{\cos\theta}{\cos(\theta+\alpha)} > h,$$

or if
$$\sqrt[3]{\frac{\sigma}{\rho}} > \frac{\cos\alpha\cos(\theta+\alpha)}{\cos\theta} \cdot \left\{\frac{\cos(\theta-\alpha)}{\cos(\theta+\alpha)}\right\}^{\frac{1}{2}}.$$

Supposing θ indefinitely small, we obtain the condition of stability for an infinitesimal displacement,

$$\sqrt[3]{\frac{\sigma}{\rho}} > \cos^2 \alpha \text{; as before, Ex. 3, Art. 68.}$$

Let the equilibrium of the cone be neutral for small displacements, that is, let

$$\sigma = \rho \cos^6 \alpha,$$

then, after a finite displacement, the action of the fluid will tend to restore the cone to its original position, if

$$\cos\alpha \cdot \cos\theta > \sqrt{\{\cos(\theta+\alpha) \cdot \cos(\theta-\alpha)\}},$$

a condition which is always true, α and θ being each less than a right angle.

In the case of neutral equilibrium of a cone, the equilibrium may therefore be characterised as stable for any finite displacement.

70. When liquid is contained in a vessel, which is slightly displaced from its original position, the preceding investigations enable us to determine the line of action of the resultant *downward* pressure.

The problem in fact in this case, as in the previous one, is the following.

A given volume, the centroid of which is H, is cut from a solid ABC by a plane, and the line CH is perpendicular to the plane;

the same volume being cut off by a plane making a very small angle with the plane AB, to determine the position of the straight line perpendicular to the second plane, and passing through the centroid of the volume cut off by it.

If the interior surface of the vessel is symmetrical with respect to the plane through H perpendicular to the line of intersection of the two planes, the line whose position is required will intersect CH in a point M, the *metacentre*, the position of which is determined by our previous results.

71. Vessel containing liquid. *A hollow vessel containing liquid, floats in liquid; required to determine the nature of the equilibrium, supposing that the body is symmetrical with respect to the vertical plane of displacement through its centre of mass, and that the centres of mass of the body and of the liquid are in the same vertical line.*

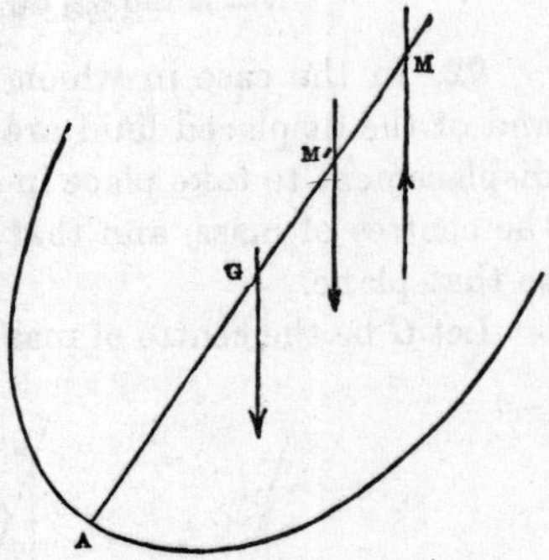

Let M be the metacentre for the displaced fluid, and M' for the contained fluid, W, W', the weights of the displaced and contained fluid.*

Taking moments about G, the centre of mass of the vessel, the resultant fluid pressures will tend to restore equilibrium, or the reverse, according as

$$W \, . \, GM - W' \, . \, GM'$$

is positive or negative, *i.e.* as

$$\frac{W}{W'} > \text{or} < \frac{GM'}{GM}.$$

EXAMPLE. *A hollow cone containing water floats in water with its axis vertical.*

Let h = the length of the axis of the cone,

h' = the length of the axis in the contained fluid,

z = the length beneath the surface of the external fluid.

---

* This is the case of a leaky ship rolling; the next article discusses the pitching of a leaky ship.

Taking 2α as the vertical angle of the cone, we have

$$HM=\tfrac{3}{4}z\tan^2\alpha.$$

But $$HG=\tfrac{2}{3}h-\tfrac{3}{4}z;$$

$$\therefore GM=\tfrac{3}{4}z\sec^2\alpha-\tfrac{2}{3}h.$$

Similarly $$GM'=\tfrac{3}{4}h'\sec^2\alpha-\tfrac{2}{3}h.$$

also $$\frac{W}{W'}=\frac{z^3}{h'^3};$$

therefore the equilibrium is stable if

$$\left(\frac{z}{h'}\right)^3>\frac{9h'\sec^2\alpha-8h}{9z\sec^2\alpha-8h},$$

z being given by the equation

$$W-W'=\tfrac{1}{3}g\rho\pi\tan^2\alpha(z^3-h'^3)=\text{weight of cone.}$$

72. In the case in which the centres of mass of the contained and of the displaced fluid are not in the same vertical, suppose the displacement to take place in direction of the vertical plane through the centres of mass, and that the body is symmetrical with respect to that plane.

Let G be the centre of mass of the body, H of the fluid displaced, H' of the contained fluid, and M, M', the metacentres.

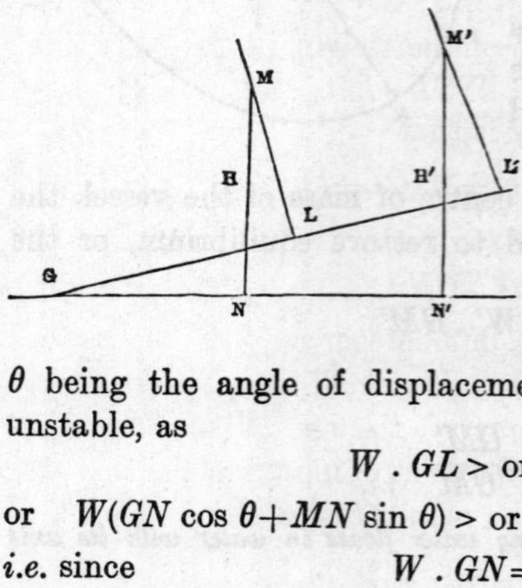

Also let GNN' be horizontal in the position of equilibrium, and GLL' the horizontal line through G in the displaced position.

Then W, W', having the same meanings as before, and θ being the angle of displacement, the equilibrium is stable or unstable, as

$$W\,.\,GL > \text{ or } < W'\,.\,GL',$$

or $$W(GN\cos\theta+MN\sin\theta) > \text{ or } < W'(GN'\cos\theta+M'N'\sin\theta),$$

i.e. since $$W\,.\,GN=W'\,.\,GN',$$

as $$\frac{W}{W'} > \text{ or } < \frac{M'N'}{MN}.$$

73. Constraints. *Stability of the equilibrium of bodies floating under constraint.*

Consider the case in which a body is free to turn about a horizontal axis fixed at a depth h. Draw GO at right angles to the axis, and, if the centre of buoyancy is not in the vertical plane through GO, let H be its projection on this vertical plane. Let C, L, N be the projections on the plane of flotation of O, G, H. Take an axis of y through C parallel to the axis of rotation and CLN as axis of x. Then if the body turns through a small angle θ about the given axis so that G, H are displaced to G', H', the vertical displacement of C is of order θ^2, and it is easy to see that the restorative moment due to the change in the displaced liquid is $g\rho Ak^2\theta$, correct to the first power of θ; where Ak^2 is the moment of inertia of the surface section about Cy. Also the loss of moment due to the displacement of H is

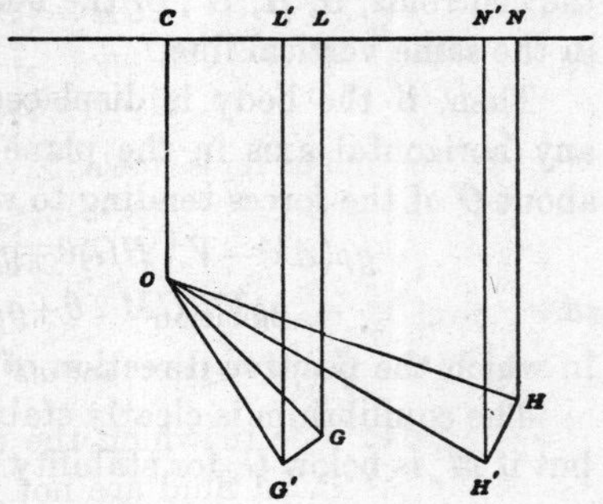

$$g\rho V\,.\,NN' = g\rho V\,.\,HH' \sin N'H'N = g\rho V(HN-h)\theta.$$

Similarly there is a loss of moment of the weight of the body due to the displacement of G of amount $W(GL-h)\theta$.

Hence the condition for stability is that

$$g\rho Ak^2 - g\rho V(HN-h) + W(GL-h)\theta$$

must be positive, with the condition

$$W\,.\,CL = g\rho V\,.\,CN.$$

Cor. If a body, floating freely in homogeneous liquid, has a plane of symmetry and is turned through a small angle θ about *any* horizontal axis in the plane of symmetry, the restorative couple is $g\rho\theta(Ak^2 - V\,.\,HG)$, where Ak^2 is the moment of inertia of the surface section about its intersection with the plane of symmetry.

74. *The equilibrium of a body floating partially immersed in two liquids.*

Let ρ be the density of the upper liquid, and $\rho+\rho'$ the density of the lower liquid.

Also let V be the total volume immersed and V' the portion of V immersed in the lower liquid, and let A, A' be the areas of the two planes of flotation. Then the forces which support the weight of the body are the weights of the masses of liquid ρV and $\rho' V'$, supposed to act upwards.

Take the case in which the body is symmetrical with regard to a vertical plane perpendicular to the plane of displacement, so that the centroids, G, H, H', of the body and of the masses ρV, $\rho' V'$ are in the same vertical line.

Then, if the body is displaced through a small angle θ about any horizontal axis in the plane of symmetry, the total moment about G of the forces tending to restore equilibrium is

$$g\rho(Ak^2 - V \,.\, HG)\theta + g\rho'(A'k'^2 - V' \,.\, H'G)\,\theta,$$

or

$$g\rho V \,.\, GM \,.\, \theta + g\rho' V' \,.\, GM' \,.\, \theta,$$

in which the positive direction of GM, GM' is upwards.

The equilibrium is clearly stable if M and M' are both above G; but if M' is below G, for stability we must have

$$\rho V \,.\, GM > \rho' V' \,.\, M'G,$$

or

$$\rho(Ak^2 - V \,.\, HG) > \rho'(V \,.\, H'G - A'k'^2).$$

75. Heterogeneous liquid.

The metacentric height in the case of heterogeneous liquid may be investigated by the method used for homogeneous liquid at the beginning of this chapter. Using the figures and notation of Art. 57, let $\rho = f(z)$ denote the density at depth z, where z is measured vertically. After the displacement the density at the point (x, y, z) of the liquid displaced is $f(z + x\theta)$, or $\rho + x\theta\dfrac{d\rho}{dz}$, to the first power of θ. The condition that the mass displaced remains constant is

$$\iiint\left(\rho + x\theta\frac{d\rho}{dz}\right)dxdydz + \iint \rho_1 x\theta dxdy = \iiint \rho dxdydz \quad . \quad (1)$$

where the volume integrals are taken through the original volume displaced, and the surface integral over the surface section refers to the wedges at the surface, and ρ_1 is the value of ρ at the surface. This condition will be satisfied if at all levels $\iint x dxdy = 0$, *i.e.* if the centroids of all horizontal sections in the original position are in the plane yz.

Again, if we suppose that the mass M_0 of liquid displaced is constant, the co-ordinates of the centres of buoyancy H, H' in the two positions are given by

$$M_0\bar{x} = \iiint \rho x dxdydz, \quad M_0\bar{y} = \iiint \rho y dxdydz, \quad M_0\bar{z} = \iiint \rho z dxdydz\,;$$

$$M_0\bar{x}' = \iiint\left(\rho + x\theta\frac{d\rho}{dz}\right)x dxdydz + \iint \rho_1 x^2\theta dxdy,$$

$$M_0\bar{y}'=\iiint\left(\rho+x\theta\frac{d\rho}{dz}\right)ydxdydz+\iint\rho_1 yx\theta dxdy,$$

$$M_0\bar{z}'=\iiint\left(\rho+x\theta\frac{d\rho}{dz}\right)zdxdydz$$

to the first order of θ.

The condition that the vertical through H' may intersect HG is $\bar{y}'=\bar{y}$, as in Art. 57,

or
$$\iiint xy\frac{d\rho}{dz}dxdydz+\iint\rho_1 xydxdy=0 \quad . \quad . \quad (2)$$

which is satisfied if at all depths the plane yz meets the horizontal section in a principal axis of that section. When conditions (1) and (2) are both satisfied we have

$$HM\,.\,\theta=HH'=\bar{x}'-\bar{x}$$
$$=\left\{\iiint x^2\theta\frac{d\rho}{dz}dxdydz+\iint\rho_1 x^2\theta dxdy\right\}\Big/M_0.$$

And if Ak^2 denotes the moment of inertia of the section at depth z about its axis in the yz plane, this gives

$$HM=\left\{\int Ak^2\frac{d\rho}{dz}dz+\rho_1 A_1 k_1^2\right\}\Big/M_0,$$

or, integrating by parts,

$$HM=\left\{[\rho Ak^2]_1^2-\int_1^2\rho\frac{d}{dz}(Ak^2)dz+\rho_1 A_1 k_1^2\right\}\Big/M_0$$
$$=\left\{\rho_2 A_2 k_2^2-\int_1^2\rho\frac{d}{dz}(Ak^2)dz\right\}\Big/M_0$$

where the suffixes 1, 2 refer to the top and bottom sections, and A_2 is zero unless the body has a flat bottom.

An alternative method will be given in the next Article.

76. *Surface of buoyancy for a solid floating in a liquid of variable density.*

Consider first the case of a body floating in a liquid formed of layers of different densities $\rho_1, \rho_2 \ldots \rho_n$ in descending order.

Let v_n denote the total volume of the solid immersed below the upper surface of the layer of density ρ_n.

As in Art. 65 let $z=c$ be the original water-line section, and let $z=c+lx+my$ denote the plane in a slightly displaced position, then we have

$$\begin{aligned}\{\rho_1 v_1+(\rho_2-\rho_1)v+(\rho_3-\rho_2)v_3+ \qquad +(\rho_n-\rho_{n-1})v_n\}(x-x_0)\\ =\{\rho_1 a_1+(\rho_2-\rho_1)a_2+\ldots \qquad +(\rho_n-\rho_{n-1})a_n\}l\\ +\{\rho_1 h_1+(\rho_2-\rho_1)h_2+\ldots \qquad +(\rho_n-\rho_{n-1})h_n\}m\,;\end{aligned}$$

and corresponding equations for $(y-y_0)$ and $(z-z_0)$ when (x_0, y_0, z_0), (x, y, z) are the centres of buoyancy in the two positions, and a_r, h_r, b_r denote

$$\iint x^2dxdy, \quad \iint xydxdy, \quad \iint y^2dxdy$$

taken over the corresponding section.

Proceeding to the case of a continuous fluid we get

$$M(x-x_0)=Al+Hm,$$
$$M(y-y_0)=Hl+Bm,$$

and
$$M(z-z_0)=\tfrac{1}{2}(Al^2+2Hlm+Bm^2),$$

where
$$\begin{aligned}M&=\rho_1 v_1+\int_1^n vd\rho\\ &=\rho_1 v_1+[\rho v]_1^n-\int_1^n \rho dv\\ &=\int_n^1 \rho dv,\end{aligned}$$

and
$$\begin{aligned}A&=\rho_1 a_1+\int_1^n ad\rho\\ &=\rho_1 a_1+[\rho a]_1^n-\int_1^n \rho da\\ &=\rho_n a_n+\int_n^1 \rho da,\end{aligned}$$

and a like expression for B, the suffixes 1, n referring to the top and bottom sections of the immersed solid, v_n being in this case clearly zero, and a_n is also zero except when the solid has a flat bottom.

The surface of buoyancy is obtained from three equations as in Art. 65, and, in the special case in which $H=0$, and the origin is at the equilibrium position of the centre of buoyancy, the equation becomes

$$2z=Mx^2/A+My^2/B,$$

and the metacentric heights are A/M and B/M.

77. *Solid floating wholly immersed.*

In this case we have similar equations, with

$$M=\int_n^1 \rho dv, \text{ and } A=\int_1^n ad\rho \text{ or } (\rho_n a_n-\rho_1 a_1)+\int_n^1 \rho da,$$

there being no displacement of the centre of buoyancy with a solid *immersed* in *homogeneous* fluid.

78. EXAMPLES. (1) *Cone of semiangle a vertex downwards.*

If x is the distance of a section from the vertex O, we have

$$a=\tfrac{1}{4}\pi x^4 \tan^4 \alpha,$$

$$\therefore da=\pi x^3 \tan^4 \alpha dx.$$

Also $dv=\pi x^2 \tan^2 \alpha dx$, so that $da=x \tan^2 \alpha dv$,

and $A/M=\int \varrho da/\int \varrho dv=\tan^2 \alpha \int x\varrho dv/\int \varrho dv$

$$=\bar{x} \tan^2 \alpha,$$

where $\bar{x}$ is the height of the centre of buoyancy above O, and thus the height of the metacentre above O is $\bar{x} \sec^2 \alpha$.

(2) *Paraboloid of latus rectum* l_0, *vertex downwards.*

Here $a=\tfrac{1}{4}\pi l_0{}^2 x^2$, $\therefore da=\tfrac{1}{2}\pi l_0{}^2 x dx$.

Also $dv=\pi l_0 x dx$, so that $da=\tfrac{1}{2}l_0 dv$,

and $A/M=\int \varrho da/\int \varrho dv=\tfrac{1}{2}l_0$.

(3) *Cylinder with axis vertical.*

Here a=constant, so that $A/M=\varrho_n a_n/M$.

79. Potential Energy. The theory of the stability of the equilibrium of floating bodies may also be based on the principle of energy and the subject may be treated from this point of view by direct calculation of the changes in the potential energy.

To find the work done in inserting a body in a sea of heavy liquid; neglecting the alteration in the level of the liquid, and the disturbance caused by the insertion of the body.

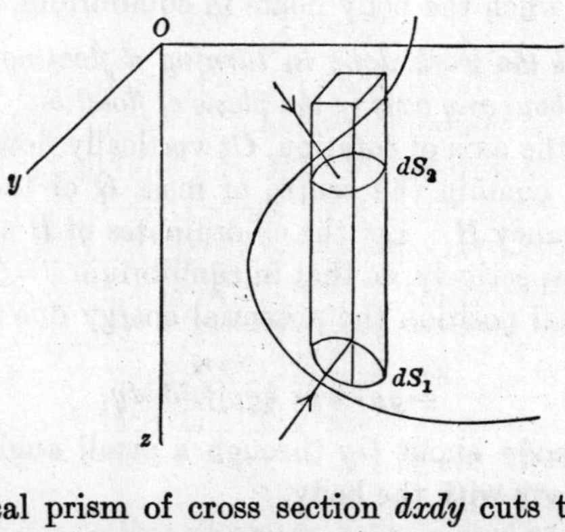

If a vertical prism of cross section $dxdy$ cuts the boundary of the body in contact with the liquid in elements dS_1, dS_2, at depths z_1, z_2, at which the pressures are p_1, p_2 respectively, and θ_1, θ_2 are the acute angles which the normals to dS_1, dS_2 make with the vertical; then the work done against the thrusts on these elements, as the depth is increased by a small amount dz, is

$$(p_1 dS_1 \cos\theta_1 - p_2 dS_2 \cos\theta_2)dz = (p_1 - p_2)dxdydz.$$

Therefore the work done in placing the body in the position under consideration

$$= \Sigma\left\{dxdy\left(\int_0^{z_1} p_1 dz - \int_0^{z_2} p_2 dz\right)\right\}$$

$$= \Sigma\left\{dxdy\int_{z_2}^{z_1} pdz\right\}$$

$$= \iiint pdxdydz \quad \text{.........................(1)},$$

where the integration extends to the volume immersed.

If the liquid be homogeneous $p = g\rho z$ and the work done

$$= g\rho\iiint zdxdydz$$

$$= g\rho V\bar{z},$$

where V is the volume of liquid displaced, and $\bar{z}$ the depth of its centroid.

When a body floats in a liquid it possesses potential energy in virtue of the work that has been done in placing it in the liquid; and if the liquid be homogeneous, and G, H the centres of mass of the body and of the liquid displaced, and ζ and $\bar{z}$ their depths, the measure of the potential energy of the body may be taken to be $g\rho V(\bar{z} - \zeta)$, or, when the body floats in equilibrium, $g\rho V \,.\, HG$.*

80. *To find the work done in turning a floating body through a small angle θ about any axis in the plane of flotation.*

Let Oy be the axis of rotation, Oz vertically downwards, and let the plane xOz contain the centre of mass G of the body and the centre of buoyancy H. Let the co-ordinates of H and G be $(\bar{x}, 0, \bar{z})$ and $(\xi, 0, \zeta)$ respectively, so that in equilibrium $\bar{x} = \xi$.

In the initial position the potential energy due to the displaced liquid

$$= g\rho V\bar{z} \text{ or } \tfrac{1}{2}g\rho\iint z^2 dxdy.$$

Turn the body about Oy through a small angle θ and let the axes Ox, Oz move with the body.

* The zero configuration is a hypothetical one, in which the space occupied by the body in the liquid is filled with liquid of the same kind, and the whole mass of the body is at the level of the free surface of the liquid.

The length to the surface of the prism of cross-section $dxdy$ immersed in the liquid becomes $z+x\tan\theta=z+x\theta$, and the depth

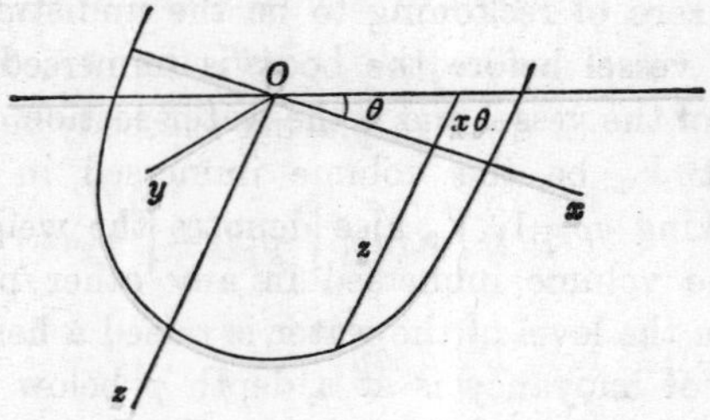

of its centre of mass is $\frac{1}{2}(z+x\theta)\cos\theta$; therefore the increase in the potential energy due to the displaced liquid

$$=\tfrac{1}{2}g\rho\iint(z+x\theta)^2(1-\tfrac{1}{2}\theta^2)dxdy-\tfrac{1}{2}g\rho\iint z^2dxdy$$

$$=\tfrac{1}{2}g\rho\theta^2\iint(x^2-\tfrac{1}{2}z^2)dxdy+g\rho\theta\iint xzdxdy.$$

But the loss of potential energy due to displacement of the body

$$=g\rho V(\zeta\cos\theta+\xi\sin\theta-\zeta)=-\tfrac{1}{2}g\rho\theta^2V\zeta+g\rho\theta V\xi,$$

therefore the total gain in potential energy is

$$E=\tfrac{1}{2}g\rho\theta^2\iint(x^2-\tfrac{1}{2}z^2)dxdy+\tfrac{1}{2}g\rho\theta^2V\zeta$$

$$=\tfrac{1}{2}g\rho\theta^2(Ak^2-V\bar{z}+V\zeta)$$

$$=\tfrac{1}{2}g\rho\theta^2(Ak^2-V\,.\,HG)\ldots\ldots\ldots\ldots\ldots\ldots(1),$$

where A is the area of the surface section of the body and k is its radius of gyration about Oy.

From this it follows that the equilibrium is stable if $Ak^2>V\,.\,HG$, and that the restorative couple is

$$\frac{dE}{d\theta}=g\rho\theta(Ak^2-V\,.\,HG).$$

The conditions previously obtained for the stability of a body floating under constraint and of a body floating in heterogeneous liquid may also be found by evaluating the changes in potential energy as far as the second power of θ. The work is to be found in earlier editions of this book, but is not regarded as of sufficient importance to be reproduced.

81. *Potential energy where a body floats in liquid contained in a cylindrical vessel.*

Take the zero of reckoning to be the undisturbed level of the liquid in the vessel before the body is immersed. Let B be the cross-section of the vessel and S the water-section of the body when floating. Let V_0 be the volume immersed in the equilibrium position; taking $g\rho=1$, V_0 also denotes the weight of the body. Let V be the volume immersed in any other position. In this latter position the level of the water is raised a height V/B, so that if the centre of buoyancy is at a depth p below the zero level, a weight V has been raised a height $p+V/2B$ and the work done is $Vp+V^2/2B$. Hence if q denote the height of the centre of gravity of the body above the zero level, the whole potential energy is

$$V_0q+Vp+V^2/2B.$$

Now let $V=V_0+v$, and let p_0 be the depth of the centroid of the volume V_0 of the body in the displaced position, so that $Vp=V_0p_0+v\xi$ where, provided that v is small, $\xi=v/2S-V/B$.

Then the potential energy is

$$V_0(q+p_0)+v\left(\frac{v}{2S}-\frac{V}{B}\right)+\frac{V^2}{2B}$$

$$=V_0(q+p_0)+v\left(\frac{v}{2S}-\frac{V_0+v}{B}\right)+\frac{(V_0+v)^2}{2B}$$

$$=V_0\zeta+\tfrac{1}{2}v^2\left(\frac{1}{S}-\frac{1}{B}\right)+\text{constant},$$

where ζ denotes the vertical distance between the centre of buoyancy and the centre of gravity.

82. EXAMPLE. *A cylinder floating in a cylinder.*

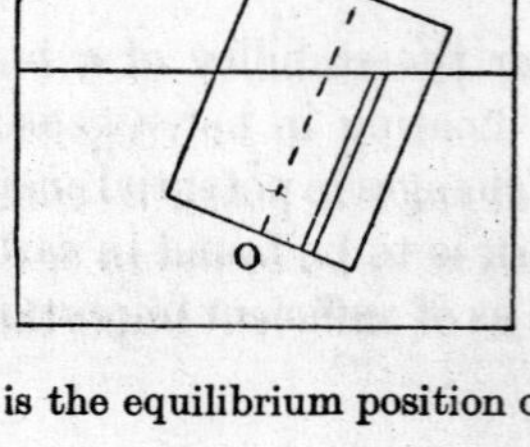

Take the origin O at the centroid of the base of the floating cylinder, which is of area A. Let the plane of the surface of the liquid be

$$lx+my+nz=p,$$

where l, m, n are direction cosines of the upward vertical.

Then $V_0=Ap/n$, and the projection on the upward vertical of the line OH_0, where H_0 is the equilibrium position of the centre of buoyancy, is

$$\frac{1}{V_0}\iint(lx+my+\tfrac{1}{2}nz)zdxdy$$
$$=\frac{1}{V_0}\iint\tfrac{1}{2}(p+lx+my)\frac{1}{n}(p-lx-my)dxdy$$
$$=\frac{1}{2nV_0}\iint\{p^2-(lx+my)^2\}dxdy$$
$$=\frac{1}{2nV_0}\{Ap^2-(\alpha l^2+\beta m^2+2\gamma lm)\};$$

where $\alpha=\iint x^2dxdy$, $\beta=\iint y^2dxdy$, $\gamma=\iint xydxdy$ integrated over the cross-section.

Also, if a, b, c are the co-ordinates of the centre of gravity G of the body, we see that

$$V_0\zeta=V_0(la+mb+nc)-\frac{1}{2n}\{Ap^2-(\alpha l^2+\beta m^2+2\gamma lm)\}$$

and $S=A/n$, so that the potential energy is

$$\tfrac{1}{2}v^2\left(\frac{n}{A}-\frac{1}{B}\right)+V_0(la+mb+nc)+\frac{1}{2n}(\alpha l^2+\beta m^2+2\gamma lm)-\tfrac{1}{2}\frac{nV_0^2}{A}+\text{const.}$$

Suppose, for example, that $a=b=0$, so that G is on the line of centroids Oz, and write $V_0=Ah$ so that h is the draught in the vertical position; then the potential energy is

$$\tfrac{1}{2}v^2\left(\frac{n}{A}-\frac{1}{B}\right)+\tfrac{1}{2}nAh(2c-h)+\frac{1}{2n}(\alpha l^2+\beta m^2+2\gamma lm).$$

In the case in which the cylinder is nearly vertical we put $n=1-\frac{1}{2}(l^2+m^2)$ approximately, and the coefficients of l^2 and m^2 become

$$\tfrac{1}{2}\{\alpha-\tfrac{1}{2}Ah(2c-h)\}\text{ and }\tfrac{1}{2}\{\beta-\tfrac{1}{2}Ah(2c-h)\},$$

so that for stability we must have $\frac{1}{2}Ah(2c-h)$ less than the least moment of inertia of the section.

If, further, the section is a circle or any form for which $\alpha=\beta$, $\gamma=0$, then the potential energy in a position in which the axis makes an angle θ with the vertical is

$$\tfrac{1}{2}v^2\left(\frac{\cos\theta}{A}-\frac{1}{B}\right)+\tfrac{1}{2}\cos\theta Ah(2c-h)+\tfrac{1}{2}\alpha\frac{\sin^2\theta}{\cos\theta}.$$

Taking the volume displaced as constant, we put $v=0$, so that for equilibrium in an oblique position we must have

$$-Ah(2c-h)+\alpha(2+\tan^2\theta)=0,$$

which gives a real value for θ, when

$$\tfrac{1}{2}Ah(2c-h)>\alpha,$$

i.e. when the vertical position is unstable.

EXAMPLES

1. If a solid paraboloid, bounded by a plane perpendicular to its axis, float with its axis vertical and vertex immersed, the height of the metacentre above the centre of gravity of the displaced liquid is equal to half the latus rectum.

2. A cone, whose vertical angle is 60°, floats in water with its axis vertical and vertex downwards; show that its metacentre lies in the plane of flotation; and that its equilibrium will be stable provided its specific gravity$>\frac{27}{64}$.

3. An isosceles wedge floats with its base horizontal, and its edge immersed; show that the equilibrium is stable for displacement in a plane perpendicular to the edge, if the ratio of the density of the wedge to that of the fluid is greater than the ratio $\cos^4 \alpha : 1$; 2α being the angle of the wedge.

4. A closed cylindrical vessel, quarter-filled with ice, is placed floating in water with its axis vertical; the weight of the vessel is one-fourth of the weight of the water which it can contain; examine the nature of the equilibrium before and after the ice melts, neglecting the change of volume consequent on the change of temperature.

5. A solid in the shape of a double cone bounded by two equal circular ends floats in a liquid of twice its density with its axis horizontal: prove that the equilibrium is stable or unstable according as the semivertical angle is less or greater than 60°.

6. The cross-section of a cylindrical ship is two equal arcs of equal parabolas of latus rectum l which touch at the keel, the common vertex of the two parabolas, so that the sides of the ship are concave to the water. The ship is floating upright with its keel at a depth h. Prove that the height of the metacentre above the keel is

$$h\left(\frac{3}{4}+\frac{h^2}{l^2}\right).$$

7. Find a solid of revolution such that, when a segment of it is immersed in liquid, the distance between the centre of buoyancy and the metacentre may be constant, whatever be the height of the segment.

8. Water rests upon mercury, and a cone is too heavy to rest without its vertex penetrating the mercury; find the density of the cone that the equilibrium may be stable assuming the cone to be completely immersed.

9. If the floating solid be a cylinder, with its axis vertical, the ratio of whose specific gravity to that of the fluid is σ, prove that the equilibrium will be stable, if the ratio of the radius of the base to the height be greater than $\{2\sigma(1-\sigma)\}^{\frac{1}{2}}$.

10. A hemispherical shell, containing liquid, is placed on the vertex of a fixed rough sphere of twice its diameter; prove that the equilibrium will be stable or unstable, according as the weight of the shell is greater or less than twice the weight of the liquid.

11. A solid of revolution floats with its vertex downwards, determine its form when the position of the metacentre is independent of the density of the liquid.

12. A solid cone is placed in a liquid with its axis vertical, and with its vertex downwards and resting on the base of the vessel containing the liquid. If the depth of the liquid be half the height of the cone, and its density four times the density of the cone, prove that the equilibrium will be stable if the vertical angle of the cone exceeds 120°.

Replacing the solid cone by a thin conical shell of the same height, of vertical angle 60°, containing liquid, up to the level of the middle point of its axis, of half the density of the liquid outside, prove that the equilibrium will be stable if the weight of the shell be less than three-fourths of the weight of the liquid inside.

13. A cylindrical vessel, the weight of which may be neglected, contains water, and the vessel is placed on the vertex of a fixed rough sphere with the centre of its base in contact with the sphere. Find the condition of stability for infinitesimal displacements, and prove that, if the equilibrium be neutral for such displacements, it will be unstable for small finite displacements.

14. A cylindrical vessel is movable about a horizontal axis passing through its centre of gravity, and is placed so as to have its axis vertical; if water be poured in, show that the equilibrium is at first unstable; and find the condition which must be satisfied, in order that it may be possible to make the equilibrium stable by pouring in enough water.

15. A thin conical vessel of given weight is movable about a diameter of its base, which is horizontal, and is partly filled with a heavy fluid; show that the equilibrium is always stable if the semivertical angle of the cone is $< 30°$; and if it be greater than this, determine when the equilibrium is stable or unstable.

16. A paraboloidal cup, the weight of which is W, standing on a horizontal table, contains a quantity of water, the weight of which is nW; if h be the height of the centre of gravity of the cup and the contained water, the equilibrium will be stable provided the latus rectum of the parabola be

$$>2(n+1)h.$$

17. A solid cone whose axis is vertical and vertex downwards is movable about an axis coincident with a generating line; to what depth must the system be immersed in water, in order that the equilibrium of the cone may be stable?

18. Prove that the work done in turning a floating body through a small angle θ round its centre of gravity is

$$\tfrac{1}{2}g\varrho(Ak^2+Ab^2-cV)\theta^2,$$

where c is the distance between the centres of gravity of the body and the liquid displaced, and b is the horizontal distance between the centre of gravity of the body and that of the area of the plane of flotation.

19. A paraboloidal cup, whose latus rectum is $4a$ and whose centre of mass is at a distance from the vertex equal to $2a$, floats in two liquids of densities σ and ϱ $(\sigma>\varrho)$; prove that the work required to turn the body through a small angle θ about a horizontal axis is

$$\tfrac{2}{3}\pi ag\theta^2\{h^3(\sigma-\varrho)+(h+h')^3\varrho\},$$

where h, h' are the lengths of the axis immersed in the fluids.

20. A thin metal circular cylinder contains water to a depth h and floats in water with its axis vertical immersed to a depth h'. Show that the vertical position is stable if the height of the centre of gravity of the cylinder above its base is less than $\tfrac{1}{2}(h+h')$.

21. A uniform liquid of density σ_2 overlies another of greater density σ_1, and a body with a plane of symmetry floats with its plane vertical so as to be in contact with both liquids. Prove that its metacentric height from the bottom of the body is

$$\frac{(z_1V_1+\kappa_1^2A_1)(\sigma_1-\sigma_2)+(z_2V_2+\kappa_2^2A_2)\sigma_2}{V_1(\sigma_1-\sigma_2)+V_2\sigma_2},$$

when V_1 is the volume submerged in the lower liquid, z_1 the height of the centre of buoyancy of this volume above the lowest point of the body, A_1, κ_1 the area and radius of gyration of the lower "water-line"; and V_2 is the whole volume below the upper "water-line," z_2 is the height of the centre of buoyancy which this volume would have if it were submerged in a single liquid, and A_2, κ_2 refer to the upper "water-line."

22. A right-angled isosceles wedge floats vertex downwards in a fluid with its base horizontal and $\frac{1}{4}$ of its volume immersed, so that its centre of gravity and metacentre coincide. Determine whether the equilibrium is really stable or unstable.

23. A solid in the form of a paraboloid of revolution floats with its axis vertical; if the centre of inertia coincides with the metacentre, prove that the equilibrium is stable.

24. A right circular cylinder of radius a rests in a liquid with its axis vertical and a length c immersed. The density at a depth z being $\phi(z)$, show that the depth of the metacentre is

$$\frac{\int_0^c z\phi(z)dz - \frac{1}{4}a^2\phi(c)}{\int_0^c \phi(z)dz}.$$

25. A paraboloid of revolution floats with its axis vertical and vertex downwards in a liquid, the density of which varies as the depth; the equilibrium will be stable or unstable, according as $4c$ is less or greater than $3(m+a)$, where c is the length of the axis, a the length immersed, and m the latus rectum of the generating parabola.

26. An oblate spheroid floats half immersed, with its axis vertical, in a liquid, the density of which varies as the square of the depth; prove that the height of the metacentre above the surface is

$$\frac{5}{8}\frac{a^2-b^2}{b}.$$

27. A solid paraboloid of revolution floats with its axis vertical, vertex downwards, and focus in the surface of a liquid, the density of which at the depth z is $\mu(a+z)$, $4a$ being the latus rectum of the generating parabola; prove that the distance of the metacentre from the vertex is $\frac{21}{8}a$.

28. A right circular solid cone of semivertical angle α floats, wholly immersed, with its vertex upwards and axis vertical, in a liquid the density of which varies as the depth. If h is the height of the cone, and b the depth of its vertex below the surface, the distance of the metacentre from the vertex is equal to

$$\frac{3}{5}h \cdot \frac{5b+4h-h\tan^2\alpha}{4b+3h}.$$

29. A cylindrical tub of sheet iron of uniform thickness, of radius a feet and weight w pounds, floats upright in water; show that its centre of gravity cannot be higher above the lower end than

$$\frac{w}{393a^2}+\frac{49a^4}{w}.$$

Prove also that, whatever be its weight, its metacentre is always more than $\cdot 7a$ feet above the lower end.

30. A cylindrical cup is made of thin uniform sheet-metal; the cup has a circular section, a flat base and an open top; its length is $4\frac{1}{2}$ times the radius of the base, and the weight of water which would fill the cup is W. Prove that the cup cannot float in water in stable equilibrium with its generators vertical, if its weight is between $(\cdot 029)\ W$ and $(\cdot 871)\ W$.

If the weight of the cup is $\frac{3}{8}W$, it can be steadied by pouring in water, so as to float with its generators vertical, provided that the weight of the water poured in lies between $\frac{1}{4}W$ and $\frac{3}{8}W$.

31. A twin steamer is formed of two equal and similar ships united alongside one another and similarly loaded. Show that, if d is the height of the metacentre above the centre of gravity in the case of the separate ships for rolling, the height in the twin ship is $d+b^2A/V$, where A is the area of the plane of flotation, V the volume immersed of either, and $2b$ the distance between the medial planes.

32. Prove that the equilibrium of a prismatic body with vertical sides near the water-line, which is so loaded that its centre of gravity coincides with its metacentre for displacement by rotation about a line parallel to its edges, is stable.

33. A cylindrical water-tank is free to swing on a horizontal axis which is a diameter of one of its cross-sections, situated below the middle of its height. Show that it will hold less water before it tips over, if the surface of the water is free, than if it is held by a lid fixed to the tank. If in the former case the water may rise to a height H above the axis of free rotation, show that in the latter it may rise an additional height $(H^2+2k^2)^{\frac{1}{2}}-H$, where the moment of inertia of the cross-section, of area A, with respect to the axis of rotation, is Ak^2.

34. A uniform right circular cylinder of height h, radius a, and specific gravity $s(<1)$ is placed with one of the circular ends below the surface of a large sheet of water; the volume of water displaced is πa^2x and the axis of the cylinder makes an angle θ with the vertical. Prove that the potential energy of the system is equal to

$$\tfrac{1}{2}w\pi a^2\{(x^2-2hsx+h^2s)\cos\theta+\tfrac{1}{4}a^2\sin\theta\tan\theta\},$$

where w is the weight of unit volume of water.

Apply this to show that, if an oblique position of equilibrium does exist with one circular end above and one below the water surface, it is a stable position.

35. Prove that a ship after passing from fresh to salt water has, in addition to change of draught, a very slight change of trim (measured by change of difference of draught fore and aft); calculate the amount in inches for a ship 300 feet long, longitudinal metacentric height 350 feet, distance of centre of gravity of area of water section from vertical through centre of gravity of ship 10 feet, increase of density $\frac{1}{40}$th part.

36. Assuming the stability of a floating body for a certain type of displacement to be measured by the height of the corresponding metacentre above the centre of gravity, show that, if a wall-sided ship is moving slowly from fresh water into salt water, this stability increases at a rate proportional to the height of the metacentre above the plane of flotation and to the rate of increase of the logarithm of the density of the water.

37. Show that, if the position of a floating body be unstable, the centre of gravity being over both metacentres, the fixing of a line in the body in the plane of the water surface gives a stable position for rotation about the line if the line lie outside a definite ellipse.

38. A heavy homogeneous cube is completely immersed with two faces horizontal in a fluid whose density$=\kappa$ times the cube of the depth. Prove that the metacentric height is $\dfrac{\kappa a^7}{120M}$, where M is the mass and a the length of an edge of the cube.

39. A thin vessel in the form of a right circular cone, whose weight is negligible, floats with axis vertical in liquid whose density is $\mu(a+z)$, z being the depth below the surface and h the length of the axis immersed. Prove that, if it contain liquid of density $\mu'\left(a+\frac{h}{4}\right)$, the equilibrium will be stable provided

$$\frac{4}{5}\left(\frac{\mu'}{\mu}\right)^{\frac{1}{3}}>\frac{4a+h}{5a+h}.$$

40. A cube, whose edge is a, floats with two faces horizontal, a length l of the vertical edges being under water. Show that the work done in turning the cube through a finite angle θ about an axis parallel to one of the horizontal edges without altering the volume of water displaced or immersing any part of the upper face of the cube is

$$W\left[\frac{a^2}{24l}\sin\theta\tan\theta-(a-l)\sin^2\frac{\theta}{2}\right],$$

where W is the weight of the cube. (See Art. 82.)

41. A ship contains water in its hold and floats in the sea. A solid is held partially immersed in the hold by a machine on land, so as to displace a weight w of water; it is then depressed so that a small extra length δx is immersed. Prove that the gain in the potential energy of the ship and contained water is

$$\left\{w-A\left(\frac{w}{B}+\frac{W}{C}\right)\right\}\delta x,$$

where W is the weight of the ship and the contained water, A is the area of the water section of the held solid, C is that of the ship, and B is the area of the surface of the contained water.

42. Show how to determine the effect on the trim of a ship of the displacement of a weight small compared to the total weight: prove that, if the displacement be across the horizontal deck in a direction making an angle θ with the medial line, the resulting slope of the deck is such that the line of greatest slope makes an angle $\tan^{-1}(m\tan\theta)$ with the medial line, where m is the ratio of the metacentric heights.

43. A log of square section floats in water with the two square faces vertical and three of the edges perpendicular to them wholly immersed. Show that there are three positions of equilibrium with a given edge not immersed, provided the specific gravity of the substance of the log lies between 23/32 and 3/4; and that if this condition be satisfied the two unsymmetrical positions are stable for rolling displacement, and the symmetrical position is unstable.

44. A homogeneous body is floating freely in stable equilibrium. Show that, if the body be turned upside down, so as to float with the same plane of flotation in a liquid of suitable density, the equilibrium will be stable.

45. Form an estimate of the effective increase in metacentric height when a ship is steadied by a rapidly spinning flywheel.

46. A uniform solid body, in the form of the portion of the paraboloid $x^2/a^2+y^2/b^2=4z/l$ cut off by the plane $z=l$, is floating freely in a liquid with its vertex downwards. A small weight is placed at the point ξ, η on its plane base, prove that these points in the plane base which suffer no vertical displacement lie on the line whose equation is

$$\frac{\xi x}{a^2-(1-n)l^2/3}+\frac{\eta y}{b^2-(1-n)l^2/3}+n=0,$$

where n^2 is the ratio of the density of the solid to that of the liquid.

CHAPTER VI

PRESSURE OF THE ATMOSPHERE

83. If a glass tube, about three feet in length, having one end closed, be filled with mercury, and then inverted in a vessel of mercury so as to immerse its open end, it will be found that the mercury will descend in the tube, and rest with its upper surface at a height of about 29 inches above the surface of the mercury in the vessel : this experiment, first made by Torricelli, has suggested the use of the *Barometer*, for the purpose of measuring the atmospheric pressure.

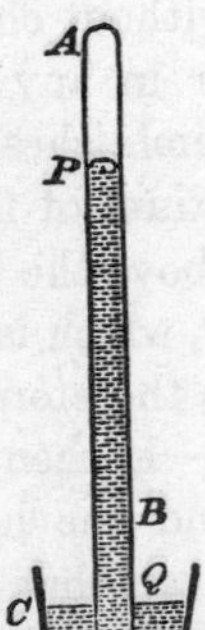

The *Barometer*, in its simplest form, is a straight glass tube *AB*, containing mercury, and having its lower end immersed in a small cistern of mercury; the end *A* is hermetically sealed, and there is no air in the branch *AB*.

It is found that the height of the surface *P* of the mercury above the surface *C* is about 29 inches, and, as there is no pressure on the surface *P*, it is clear that the pressure of the air on *C* is the force which sustains the column of mercury *PQ*.

We have shown that the pressure of a fluid at rest is the same at all points of the same horizontal plane ; hence the pressure at *C* is equal to the pressure of the mercury at *Q*.

Let σ be the density of mercury, and Π the atmospheric pressure at *C*, then

$$\Pi = g\sigma PQ,$$

and the height PQ measures the atmospheric pressure.

On account of its great density, mercury is the most convenient fluid which can be employed in the construction of barometers, but the pressure of the air may be measured by using any kind of liquid. The density of mercury is about 13·568 times that of water, and therefore the height of the column of water in the water-barometer would be about $33\frac{3}{4}$ feet.

The density of mercury changes with the temperature, and σ must therefore be expressed as a function of the temperature.

Experiment shows that, for an increase of 1° centigrade, the expansion of mercury is $\frac{1}{5550}$th of its volume; hence if σ_t be the density at a temperature $t°$, and σ_0 at a temperature 0°,

$$\sigma_0=\sigma_t\left(1+\frac{t}{5550}\right)=\sigma_t(1+\cdot00018018t);$$

$$\therefore\ \sigma_t=\sigma_0(1-\theta t) \text{ if } \theta=\cdot00018018,$$

and $$\Pi=g\sigma_0(1-\theta t)PQ.$$

By means of the formula, $\Pi=g\sigma_0(1-\theta t)h$, the atmospheric pressure at any place can be calculated, making due allowance for the change in the value of g consequent on a change of latitude. It is found that this pressure is variable at the same place, with or without changes of temperature, and that in ascending mountains, or in any way rising above the level of the place, the pressure diminishes. This is in accordance with the theory of the equilibrium of fluids, for, in ascending, the height of the column of air above the barometer is diminished, and the pressure of the air upon C, which is equal to the weight of the superincumbent column of air, is therefore diminished, and the mercury must descend in the tube.

If then a relation be found between the height of the mercury and the height through which an ascent has been made, it is clear that by observations, at the *same* time, of the barometric columns at two stations, we shall be able to determine the difference of their altitudes.

We shall investigate a formula for this purpose; but it is first necessary to state the laws which regulate the pressures of the air and gases at different temperatures, and also the laws of the mixture of gases.

84. We have before stated the relation

$$p=k\rho(1+at)$$

between the pressure, density, and temperature of an elastic fluid: it is deduced from the two following results of experiment:

(1) *If the temperature be constant, the pressure of air varies inversely as its volume.* (*Boyle's Law.*)

(2) *If the pressure remain constant, an increase of temperature*

of 1° C. *produces in a mass of air an expansion* ·003665 *of its volume at* 0° C. (*Dalton's and Gay-Lussac's Law.*)

Hence, if p be the pressure and ρ_0 the density of air, at a temperature zero,

$$p=k\rho_0.$$

Suppose now the temperature increased to t, the pressure remaining the same: the conception of this may be assisted by considering the air to be contained in a cylinder in which a movable piston fits closely, and has applied to it a constant force, so that an increase of the elastic force of the air would have the effect of pushing out the piston, until the equilibrium is restored by the diminution of density, and consequent diminution of pressure: we shall then have from the 2nd law,

$$\rho_0=\rho(1+\alpha t),$$

taking ρ as the new density and $\alpha=\cdot003665$;

$$\therefore p=k\rho(1+\alpha t).$$

If p', ρ' be the pressure and density of the same fluid at a temperature t',

$$p'=k\rho'(1+\alpha t'),$$

and

$$\frac{p}{p'}=\frac{\rho}{\rho'}\frac{1+\alpha t}{1+\alpha t'}.$$

The quantity α is very nearly the same for gases of all kinds, but k has different values for different gases, and must of course be determined experimentally in every case.

85. Absolute Temperature. If we imagine the temperature of a gas lowered until its pressure vanishes, without any change of volume, we arrive at what is called the absolute zero of temperature, and absolute temperature is measured from this point.

Assuming t_0 to represent this temperature on the centigrade thermometer, we obtain, from the equation $1+\alpha t_0=0$,

$$t_0=-\frac{1}{\alpha}=-273°.$$

In Fahrenheit's scale the reading for absolute zero is $-459°$.

The equations,

$$p=k\rho(1+\alpha t),$$
$$0=k\rho(1+\alpha t_0),$$

lead to

$$p=k\rho\alpha(t-t_0)$$
$$=k\rho\alpha T,$$

if T be the absolute temperature.

Since ρV is constant, it follows that pV/T is constant, and this law expresses, in the absolute scale, the relation between pressure, volume, and temperature.

86. Mixtures. *The pressure of a mixture of different elastic fluids.*

Consider two different gases, contained in vessels of which the volumes are V and V', and let their pressures and temperatures, p and t, be the same.

Let a communication be established between the two vessels, or transfer both the gases to a closed vessel, the volume of which is $V+V'$: it is found in the case in which no chemical action takes place, that the two gases do not remain separate, but permeate each other until they are completely mixed, and that, when equilibrium is attained, the pressure and temperature are the same as before. From this important experimental fact we can deduce the following proposition.

If two gases having the same temperature be mixed together in a vessel, the volume of which is V, *and if the pressure of the two gases, alone filling the volume* V, *be* p *and* p′, *the pressure of the mixture will be* p+p′.

Suppose the two gases separated; let the gas, of which the pressure is p, have its volume changed, without any alteration of temperature, until its pressure becomes p'; its volume will be, by Boyle's law, pV/p'.

Let the two gases be now mixed in a vessel, of which the volume is

$$V+\frac{p}{p'}V, \text{ or } \frac{p+p'}{p'}V;$$

the pressure of the mixture will still be p', and the temperature will be unaltered. If the mixture be then compressed into a volume V, its pressure will become, by the application again of Boyle's law, $p+p'$.

This result is obviously true for a mixture of any number of gases.

87. *Two volumes* V, V′ *of different gases, at pressures* p, p′ *respectively, are mixed together, so that the volume of the mixture is* U; *to find the pressure of the mixture.*

The pressures of the two gases, reduced to the volume U, are respectively

$$\frac{V}{U}p, \frac{V'}{U}p',$$

and therefore, by the preceding article, the pressure of the mixture is

$$\frac{V}{U}p+\frac{V'}{U}p';$$

and if ϖ be this pressure, we have

$$\varpi U=pV+p'V'.$$

If the absolute temperatures of the gases before mixture are T and T', and if after mixture the absolute temperature is τ, and the volume U, the pressures of the gases will be respectively

$$\frac{pV}{T}\frac{\tau}{U} \text{ and } \frac{p'V'}{T'}\frac{\tau}{U}.$$

Hence ϖ, the pressure of the mixture, is the sum of these two quantities, and therefore

$$\frac{\varpi U}{\tau}=\frac{pV}{T}+\frac{p'V'}{T'}.$$

In the case of the mixture of any number of gases, we have

$$\frac{\varpi U}{\tau}=\Sigma\frac{pV}{T}.$$

88. The laws and results of the preceding articles are equally true of vapours, the only difference between the mechanical qualities of vapours and gases, irrespective of their chemical characteristics, being that the former are easily condensed into liquid by lowering the temperature, while the latter can only be condensed by the application either of great pressure or extreme cold, or a combination of both.*

89. Vapour. If water be introduced into a space containing dry air, vapour is immediately formed, and it is found that the pressure and density of the vapour are dependent only on the temperature, and are quite independent of the density of the air, and indeed are exactly the same if the air be removed. If the

* Professor Faraday succeeded in condensing carbonic acid gas, and other gases requiring a considerable pressure for the purpose, and the result of his experiments led to the conclusion that, in all probability, all gases are the vapours of liquids. This conclusion was remarkably supported in 1877, when M. Pictet, in the early part of the year, liquefied oxygen by applying to it a pressure of 300 atmospheres, and, in December of the same year, M. Cailletet liquefied nitrogen, and atmospheric air. In 1884 hydrogen was liquefied by Wroblewski, in 1899 Dewar obtained solid hydrogen, and now liquid air and various other gases in liquid form are articles of commerce.

temperature be increased or the space enlarged, an additional quantity of vapour will be formed, but if the temperature be lowered or the space diminished, some portion of the vapour will be condensed.

While a sufficient quantity of water remains, as a source from which vapour is supplied, the space will be always saturated with vapour, that is, there will be as much vapour as the temperature admits of; but if the temperature be so raised that all the water is turned into vapour, then for that, and all higher temperatures, the pressure of the vapour will follow the same law as the pressure of the air.

In any case, whether the space be saturated or not, if p be the pressure of the air, and ϖ of the vapour, the pressure of the mixture is $p+\varpi$.

90. The atmosphere always contains aqueous vapour, the quantity being greater or less at different times; if any portion of the space occupied by the atmosphere be saturated with vapour, that is, if the density of the vapour be as great as it can be for the temperature, then any reduction of temperature will produce condensation of some portion of the vapour, but if the density of the vapour be not at its maximum for that temperature, no condensation will take place until the temperature is lowered below the point corresponding to the saturation of the space.

Formation of Dew. If any surface, in contact with the atmosphere, be cooled down below the temperature corresponding to the saturation of the space near it, condensation of the aqueous vapour will ensue, and the condensed vapour will be deposited in the form of *dew* upon the surface. The formation of dew on the ground depends therefore on the cooling of its surface, and this is in general greater and more quickly effected when the sky is free from clouds, and when, consequently, the loss of heat by radiation is greater than under other circumstances.

The **Dew Point** is the temperature at which dew first begins to be formed, and must be determined by actual observation.

The pressure of vapour corresponding to its saturating densities for different temperatures must also be determined experimentally, and, if this be effected, an observation of the dew point at once determines the pressure of the vapour in the atmosphere. For if t' be the dew point, and p' the known corresponding pressure, then at any other temperature t above t' the pressure p is given by the equation

$$\frac{p}{p'}=\frac{1+at}{1+at'}.$$

91. *Effect of compression or dilatation on the pressure and temperature of a gas.*

It is found by experiment that if a quantity of air, enclosed in a vessel impervious to heat, be compressed, its temperature is raised; and that, if a quantity of air, enclosed in any kind of vessel, be suddenly compressed, so that there is no time for the heat to escape, the temperature is similarly raised.

92. Thermal Capacity. The thermal capacity of a body is measured by the amount of heat required to raise the temperature one degree.

The unit of heat which is actually employed is the quantity of heat required to raise by one degree the temperature of one unit of mass of water, supposed to be between 0° C. and 40° C.

Specific Heat. The specific heat of a body is the thermal capacity of one unit of mass, or, which is the same thing, it is the ratio of the amount of heat required to increase by 1° the temperature of the body to the amount of heat required to increase by 1° the temperature of an equal weight of water.

If an amount of heat dQ produce in the unit of mass a change of temperature dt, the measure of the specific heat is $\frac{dQ}{dt}$.

In gases it is necessary to consider two cases: (1) when the pressure remains constant, the gas being allowed to expand, (2) when the volume remains constant.

We shall denote the specific heat in these two cases by the symbols c_p and c_v.

It is easy to see that c_p is greater than c_v, for in the first case the heat imparted does work in expanding the gas as well as in raising its temperature.

93. Internal Energy. A mass of gas in a given state possesses internal energy depending upon the configuration and motion of its molecules. The difference between the energies in two given states depends only upon those states, and not upon the mode of change from the one to the other. If we denote by U the difference between the internal energies in any assigned state and in some standard state, then dU is a perfect differential of a function determined by the state of the gas.

For a gas the first law of thermodynamics may be expressed by the relation

$$dQ = dU + pdv \quad . \quad . \quad . \quad (1)$$

or the quantity of heat imparted is equal to the increase in internal energy together with the work done by the pressure as the gas expands.

A *perfect gas* is an ideal substance which is assumed to obey the relation $pv = KT$ for all ranges of temperature, where T denotes absolute temperature and K is a constant. There are experimental reasons for concluding that for such a gas the internal energy U is a function of T only.

If we suppose the volume to be kept constant while heat is imparted, then $c_v = dQ/dT$. Hence it follows from (1) that $c_v = dU/dT$; but U is a function of T alone, therefore c_v is a function of T alone. Now it is found that for the permanent gases and for all but very high or very low temperatures, c_v is independent of T, consequently it is assumed that for a perfect gas c_v is independent of T, i.e. c_v is a constant and $dU = c_v dT$.

Hence, for a perfect gas, (1) may be written

$$dQ = c_v dT + pdv \quad . \quad . \quad . \quad . \quad (2)$$

where

$$pv = KT.$$

Therefore

$$pdv + vdp = KdT,$$

so that

$$dQ = c_v dT + KdT - vdp \quad . \quad . \quad . \quad (3)$$

Now, suppose that the pressure is kept constant while a quantity of heat dQ is imparted, so that $c_p = dQ/dT$ or $dQ = c_p dT$. Substituting this value in (3) and putting $dp = 0$, we get

$$c_p - c_v = K \quad . \quad . \quad . \quad . \quad (4)$$

Consequently c_p is also a constant for a perfect gas and, as stated in the last article, it is greater than c_v.

94. Adiabatic Expansion. Let a change of state take place without any heat being imparted to or lost from the gas. Such an expansion or compression is called an *adiabatic change.*

In this case, since no heat is supplied or lost, we have

$$dQ=0,$$

therefore, from (2)

$$0=c_v dT+pdv.$$

But $pv=KT$, and from (4) $K=c_p-c_v$,

therefore
$$pdv+vdp=(c_p-c_v)dT,$$

and eliminating dT gives

$$pdv+vdp+\left(\frac{c_p}{c_v}-1\right)pdv=0,$$

or
$$\frac{dp}{p}+\frac{c_p}{c_v}\frac{dv}{v}=0.$$

On integration we find that

$$pv^\gamma=\text{const.},$$

where γ denotes the constant ratio c_p/c_v.

The equation $pv^\gamma=$constant is, in thermodynamics, the equation of the adiabatic, or isentropic lines, and it represents the relation between the pressure and volume of a mass of gas, when, during a change of volume, no heat is lost or imparted.

The equation is true in the case of a sudden compression or dilatation of a mass of air, because there is no time for any sensible loss of heat, or for any addition of heat from external sources. It will be found that this relation is of great importance in the theory of sound.

95. *To find the work done in compressing a gas isothermally.*

Let v be the volume of a gas at the pressure p, dS an element of the surface of the vessel containing it, and dn an element of the normal to dS drawn inwards.

Then the work done in a small compression

$$=p\Sigma dSdn=-pdv,$$

and the work done in compressing from volume V to V'

$$=-\int pdv=-\int\frac{Cdv}{v}, \text{ since } pv=C,$$

$$=C\log\frac{V}{V'}=pv\log\frac{V}{V'}.$$

If the compression takes place in a vessel surrounded by the atmosphere, as for example if the gas is confined in a cylinder by a piston, the pressure of the atmosphere assists in the work of compression. Thus if the initial volume is V at atmospheric pressure Π, the external work done in compressing it to volume V'

$$=-\int_{V}^{V'}(p-\Pi)dv, \text{ where } pv=\Pi V$$

$$=\Pi V \log\frac{V}{V'}-\Pi(V-V').$$

96. *The work done during an adiabatic compression of a gas.*

In the last paragraph we have assumed that the compression is isothermal.

This state of things can be secured by performing the operation so slowly that any heat which may be generated is dissipated during the process.

If the compression is adiabatic, that is, if the process is so arranged that no heat is lost or imparted, which is practically the case when the compression is very rapid, we have from Art. 94 the relation

$$pv^{\gamma}=\text{constant}=C.$$

Hence it follows that the work done in compressing from volume V to volume U

$$=-\int pdv=-\int Cv^{-\gamma}dv$$

$$=\frac{C}{1-\gamma}(V^{1-\gamma}-U^{1-\gamma}).$$

97. Isothermal Atmosphere.

On the hypothesis of uniform temperature the law of pressure is given by

$$dp=-g\rho dz,$$

where p, ρ denote pressure and density at a height z. If p_0, ρ_0 denote the values at a height z_0, we have

$$\frac{p}{\rho}=\frac{p_0}{\rho_0}=k,$$

and $\therefore$

$$k\log p=C-gz\,;$$

whence

$$\log\frac{p}{p_0}=-\frac{g}{k}(z-z_0) \quad . \quad . \quad . \quad (1)$$

If we take $z_0=0$ and suppose H to be the height of a homogeneous atmosphere of density ρ_0, that would produce the pressure p_0, we have $p_0=g\rho_0 H$, so that $k=gH$, and $\log p/p_0=-z/H$, or

$$p=p_0e^{-z/H}.$$

This shows that, as the altitude increases in arithmetical progression, the pressure decreases in geometrical progression.

Formula (1) may be used for comparing differences of level by observing barometric pressure. Thus we have

$$z-z_0=-\frac{k}{g}\log\frac{p}{p_0},$$

on the hypothesis that the temperature is constant. If the temperature be not constant, the relation between p and ρ is $p=k\rho(1+\alpha t)$, and if τ, τ_0 be the temperatures at the two stations and we proceed on the hypothesis of a mean uniform temperature $t=\frac{1}{2}(\tau_0+\tau)$, we have

$$z-z_0=-\frac{k}{g}\{1+\tfrac{1}{2}\alpha(\tau_0+\tau)\}\log\frac{p}{p_0}.$$

This formula may be further corrected by allowing for the difference in the value of gravity at different altitudes; thus, if g is the measure of gravity at sea level and r is the earth's radius, the attractive force at a height z is measured by $gr^2/(r+z)^2$. For accurate results corrections must be made to the barometer readings so as to allow for the difference of temperature of the mercury at different levels and for the aqueous vapour in the atmosphere, but a more detailed discussion is beyond the scope of our present purpose.

98. Convective Equilibrium. An alternative hypothesis is that of the convective equilibrium of temperature in the atmosphere. As explained by Lord Kelvin,* "when all the parts of a fluid are freely interchanged and not sensibly influenced by radiation and conduction, the temperature of the fluid is said to be in a state of convective equilibrium." This state implies that if equal masses of air at different levels were interchanged without gain or loss of heat, *i.e.* adiabatically, they would merely interchange pressure, density and temperature so that on the whole there would be no change. In this case therefore the equations are

$$dp=-g\rho dz \quad . \quad . \quad . \quad . \quad (1)$$

$$p=k\rho^{\gamma} \text{ and } p=K\rho T,$$

where T denotes absolute temperature at the height z;

$$\therefore k\gamma\rho^{\gamma-2}d\rho=-gdz,$$

* *Collected Papers*, vol. iii. p. 255.

and by integration

$$\frac{k\gamma}{\gamma-1}\rho^{\gamma-1}=C-gz\,;$$

$$\therefore\ \frac{\gamma}{\gamma-1}\frac{p}{\rho}=C-gz\,;$$

$$\therefore\ \frac{\gamma}{\gamma-1}K(T-T_0)=-gz,$$

where T_0 denotes the absolute temperature at sea-level;

$$\therefore\ \frac{T}{T_0}=1-\frac{\gamma-1}{\gamma}\cdot\frac{gz}{KT_0}.$$

And if H is the height of the homogeneous atmosphere

$$K\rho_0T_0=p_0=g\rho_0H\,;$$

$$\therefore\ \frac{T}{T_0}=1-\frac{\gamma-1}{\gamma}\cdot\frac{z}{H}\quad . \quad . \quad . \quad (2)$$

If in equation (1) we take $gr^2/(r+z)^2$ instead of g, as before, we get on integration and substitution as above

$$\frac{T}{T_0}=1-\frac{\gamma-1}{\gamma}\cdot\frac{rz}{H(r+z)}\quad . \quad . \quad . \quad (3)$$

99. The following problem is illustrative of the principles of this chapter.

A piston without weight fits into a vertical cylinder, closed at its base and filled with atmospheric air, and is initially at the top of the cylinder; water being poured slowly on the top of the piston, find how much can be poured in before it will run over.

Let a be the height of the cylinder, and z the depth to which the piston will sink; then in the position of equilibrium the pressure of the air in the cylinder is $\Pi+g\varrho z$, where Π is the atmospheric pressure, and ϱ the density of water; but

$$\text{this pressure} : \Pi = a : a-z\,;$$

$$\therefore\ \frac{\Pi a}{a-z}=\Pi+g\varrho z.$$

Let h be the height of the water-barometer,

$$\therefore\ \Pi=g\varrho h,$$

$$ha=(a-z)(h+z),$$

and

$$z=0 \text{ or } a-h.$$

Unless then the height of the cylinder is greater than h, no water can be poured in, for, even if the piston be forced down and water then poured on it, the pressure of the air beneath will raise the piston.

The negative solution, when $a<h$, can however be explained as the solution

of a different problem leading to the same algebraic equation. Suppose the cylinder to be continued above the piston, and let it be required to raise the piston through a space z by a force which shall be equal to the weight of the cylindrical space z of water.

This leads to the equation

$$\frac{\Pi - g\varrho z}{\Pi} = \frac{a}{a+z},$$

or

$$z = h - a.$$

EXAMPLES

1. The readings of a perfect mercurial barometer are α and β, while the corresponding readings of a faulty one, in which there is some air, are a and b; prove that the correction to be applied to any reading c of the faulty barometer is

$$\frac{(\alpha-a)(\beta-b)(a-b)}{(a-c)(\alpha-a)-(b-c)(\beta-b)}.$$

2. If a thermometer, plunged incompletely in a liquid whose temperature is required, indicate a temperature t, and τ be that of the air, the column not immersed being m degrees, prove that the correction to be applied is

$$\frac{m(t-\tau)}{6840+\tau-m},$$

1/6840 being the expansion of mercury in glass for 1° of temperature, assuming that the temperature of the mercury in each part is that of the medium which surrounds it.

3. A closed vertical cylinder of unit sectional area contains a piston, weight W. The piston is originally halfway up the cylinder, and the space above and below is filled with saturated air. On being left to itself the piston sinks to half its former height; prove that the tension of the saturated vapour is $3W-4\Pi$ where Π is the pressure of the atmosphere: the temperature being supposed the same at the end and beginning of the process.

4. A vertical barometer tube is constructed, of which the upper portion is closed at the top, and has a sectional area a^2, the middle portion is a bulb of volume b^3, and the lower portion has a section c^2, and is open at the bottom; the mercury fills the bulb and part of the upper and lower portions of the tube, and is prevented from running out below by means of a float against which the air presses; the upper part of the tube is a vacuum: find the change of position of the upper and lower ends of the mercurial column, due to a given alteration of the pressure of the atmosphere.

Show also that, if the whole volume of the mercury in the instrument be c^2H, where H is the height of the barometer, the upper surface will be unaffected by changes of temperature.

5. A cylindrical diving-bell sinks in water until a certain portion V remains occupied by air, and in this position a quantity of air, whose volume under the atmospheric pressure was $2V$, is forced into it. Show how far the bell must sink in order that the air may occupy the same space as in the first position.

Find also the condition that when the air is forced in at the first position no air may escape from beneath the bell.

6. A vessel, in the form of the surface generated by the revolution about its axis of an arc of a parabola terminated by the vertex, is immersed, mouth

downwards, in a trough of mercury; show that the pressure of the air contained in the vessel varies inversely as the square of the distance of the vertex of the vessel from the surface of the mercury within it. Supposing the length of the axis of the vessel to be to the height of the barometer as 45 is to 64, find the depth of the surface of the mercury within the vessel, when the whole vessel is just immersed.

7. A piston without weight fits into a vertical cylinder, closed at its base and filled with air, and is initially at the top of the cylinder; if water be slowly poured on the top of the piston, show that the upper surface of the water will be lowest when the depth of the water is $\sqrt{(ah)}-h$, where h is the height of the water-barometer, and a the height of the cylinder.

8. A cylindrical well of depth h and section A is maintained at constant temperature; if ϱ_0 and ϱ_1 are the densities of the air at the top and bottom, show that the total amount of air contained is $Ah(\varrho_1-\varrho_0)/(\log \varrho_1-\log \varrho_0)$: if the barometer at the top stand at 30 inches, and at the bottom at 31 inches, show that the mean density of the air in the well will differ from that due to a pressure of 30·5 inches by about 1 part in 11,000.

9. A straight tube, closed at one end and open at the other, revolves with a constant angular velocity about an axis meeting the tube at right angles; neglecting the action of gravity, find the density of the air within the tube at any point.

10. A bent tube of uniform bore, the arms of which are at right angles, revolves with constant angular velocity ω about the axis of one of its arms, which is vertical and has its extremity immersed in water. Prove that the height to which the water will rise in the vertical arm is

$$\frac{\Pi}{g\varrho}\left(1-e^{-\frac{\omega^2 a^2}{2k}}\right),$$

a being the length of the horizontal arm, Π the atmospheric pressure, and ϱ the density of water, and k the ratio of the pressure of the atmosphere to its density.

11. A thin uniform circular tube of radius a contains air and rotates with angular velocity ω about an axis in its plane, distant c from the centre; find the pressure at any point neglecting the weight of the air. If c is less than a, and if p and p' are the greatest and least pressures, prove that

$$\log \frac{p}{p'}=\frac{\omega^2}{2k}(a+c)^2.$$

12. Two bulbs containing air are connected by a horizontal glass tube of uniform bore, and a bubble of liquid in this tube separates the air into two equal quantities. The bubble is then displaced by heating the bulbs to temperatures t degrees and t' degrees: prove that, if the temperature of each bulb be decreased τ degrees, the bubble will receive an additional displacement which bears to the original displacement the ratio of

$$2a\tau : 2+a(t+t'-2\tau),$$

where a is the coefficient of expansion.

13. A conical shell, vertical angle $\pi/2$, and height H, can hold double its own weight of water. It is inverted and immersed, axis vertical, in a mass of water. The water is now made to rotate with angular velocity $(7g^3/2H^3)^{\frac{1}{4}}$ and the cone sinks till its vertex lies in the surface: prove that the height of the water-barometer is to that of the cone as $3:\sqrt[3]{28}$.

14. If the pressure of the air varied as the $(1+1/m)^{th}$ power of the density, show that, neglecting variations of temperature and gravity, the height of the atmosphere would be equal to $(m+1)$ times the height of the homogeneous atmosphere.

15. A piston of weight w rests in a vertical cylinder of transverse section k, being supported by a depth a of air. The piston rod receives a vertical blow P, which forces the piston down through a distance h: prove that

$$(w+\Pi k)\left\{h+a\log\left(1-\frac{h}{a}\right)\right\}+\frac{gP^2}{2w}=0,$$

Π being the atmospheric pressure.

16. Prove that, if the temperature in the atmosphere fall uniformly with the height ascended, the height of a station above sea level is given by

$$z=a\{1-(h/h_0)^m\},$$

where h, h_0 are the readings of the barometer at the station and at sea level respectively, and a, m are constants.

17. Show that in an atmosphere in "convective equilibrium" the temperature would diminish upwards with a uniform gradient; and calculate this gradient in degrees centigrade per 100 metres, assuming the following data (in C.G.S. units):

height of barometer	$=76{\cdot}0$,
temperature (absolute)	$=272°$ C.,
density of air	$={\cdot}00129$,
density of mercury	$=13{\cdot}60$,
ratio of specific heats (γ)	$=1{\cdot}42$.

18. In a vertical column of perfect gas the pressure and absolute temperature at any height z are p and T. Prove that

$$z=\frac{p_0}{\varrho_0 g T_0}\int_p^{p_0}\frac{T\,dp}{p},$$

where p_0, ϱ_0, T_0 are pressure, density and absolute temperature at the bottom.

Height is measured in an aeroplane by means of a specially graduated aneroid barometer. The graduations are such that the true height would be read direct if the temperature of the atmosphere were uniformly at 10° C. Show that the instrument will read differences of height correctly whatever the barometric pressure at ground level.

To find the true height when the temperature is not uniform, it is necessary to read the temperature during the ascent. Show that the true height corresponding to a recorded height z is $\int_{z_0}^{z}\frac{T}{283}\,dz'$, where z_0 is the reading at ground level and T the absolute temperature when the reading is z'.

19. A perfectly flexible balloon contains a light gas of total mass m. At the ground level it is at the same temperature as the surrounding air. Prove that it will exert the same lift at all heights if it remains at the same temperature as the air round it, but that, if the gas inside expands adiabatically, the lift at height z will be less than the lift at the ground level by the amount

$$mg\sigma\left\{1-\left(1-\frac{z}{H}\right)^{\frac{\gamma'-\gamma}{(\gamma-1)\gamma'}}\right\},$$

where σ is the ratio of the density of air to that of the gas under standard conditions, γ, γ' are the ratio of the specific heats for air and for the gas and H is the height of the atmosphere, *i.e.* the height at which pressure, temperature, and density vanish. It is supposed that the balloon is never fully extended.

CHAPTER VII

CAPILLARITY

100. It is a well-known fact that if a glass tube of small bore be dipped in water, the water inside the tube rises to a higher level than that of the water outside.

It is equally well known that if the tube be dipped in mercury, the mercury inside is depressed to a lower level than that of the mercury outside.

If a glass tumbler contain water it will be seen that at the line of contact the surface is curved upwards and appears to cling to the glass at a definite angle.

If the tumbler be carefully filled, the level of the water will rise above the plane of the top of the tumbler, the water bulging over the round edge of the top.

If water be spilt on a table, it has a definite boundary, and the curved edges cling to the table.

These facts, and many others, are explained by the existence of forces between the molecules of the fluids, and of the solids and fluids, in contact; the field of action of the force exerted by any particular molecule being infinitely small.* And since

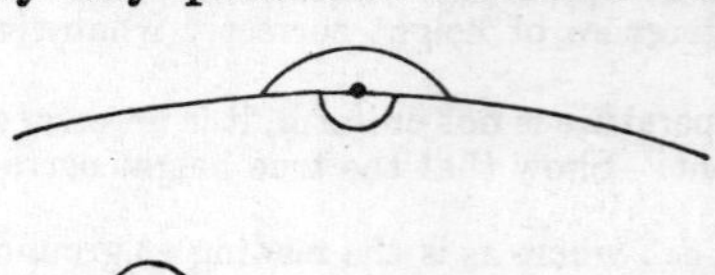

these molecular forces are only exerted at very small distances, it follows that as far as molecular forces are concerned, every element of a homogeneous body, not near its bounding surface, is under the same conditions; but that at the surface itself the sphere of action of a particular molecule is incomplete, and the molecule also falls within the field of action of molecules of whatever matter is on the other side of the bounding surface.

* The field through which capillary forces are exerted is extremely small. In Quincke's experiments the same phenomena were observed with water in a glass tube silvered with a coating ·0000542 mm. thick, as in a silver tube of the same diameter. *Pogg. Ann.*, cxxxix. (1870), p. 1.

Also if we assume that the linear dimensions of the field of action are infinitely small as compared with the radii of curvature of the surface, then all parts of the surface of separation of two homogeneous substances are under similar conditions as far as molecular forces are concerned, and the surface potential energy due to molecular forces must be in a constant ratio to the area of the surface, the constant depending on the nature of the substances in contact.

101. Surface Tension.

We shall see shortly that the surface potential energy is such as would exist if the surface were in a state of uniform tension T, so that the tension in the surface across *any* short line of length δs in the surface is $T\delta s$ at right angles to the line δs.

We proceed to show that, if such a surface tension exists, then there is a relation between the surface tension T, the curvatures of the surface, and the difference ϖ of the pressures on opposite sides of the surface.

Let the equation of the surface be $z=f(x, y)$. Consider the equilibrium of any portion S of the surface bounded by a curve s without singularities. The resultant of the tensions $T\delta s$ across all the elements δs of the curve s must balance the resultant of the pressure differences $\varpi\delta S$ on the various elements δS of the surface S.

Let λ, μ, ν denote the direction cosines of the normal to the surface at the point (x, y, z), and let l, m, n denote the direction cosines of the tension $T\delta s$ across δs. The direction cosines of the tangent to the element δs at (x, y, z) are dx/ds, dy/ds, dz/ds, or x', y', z'; and, since the tension is at right angles to δs and to the normal to the surface, therefore (l, m, n), (λ, μ, ν), and (x', y', z') are the direction cosines of three mutually perpendicular lines, and

$$\frac{l}{\mu z'-\nu y'}=\frac{m}{\nu x'-\lambda z'}=\frac{n}{\lambda y'-\mu x'}=1.$$

The equation of equilibrium obtained by resolving parallel to the axis of z is

$$\iint \varpi\nu dS-\int Tnds=0,$$

which is equivalent to

$$\iint \varpi dxdy-\int T(\lambda dy-\mu dx)=0,$$

where the integrations are over the projection of S on the xy plane and round the boundary of this projection. By using Green's

Theorem * for transforming the line integral into a surface integral, this becomes

$$\iint\left\{\varpi-T\left(\frac{\partial\lambda}{\partial x}+\frac{\partial\mu}{\partial y}\right)\right\}dxdy=0.$$

Since this integral must vanish for all such ranges of integration, the integrand must be zero. Hence we have

$$\varpi=T\left(\frac{\partial\lambda}{\partial x}+\frac{\partial\mu}{\partial y}\right).$$

Now, in the ordinary notation, with p, q, r, s, t as partial differential coefficients of z with regard to x and y, we have

$$\lambda,\ \mu,\ \nu=\frac{p,\ q,\ -1}{(p^2+q^2+1)^{\frac{1}{2}}}.$$

From this we find that

$$\frac{\partial\lambda}{\partial x}+\frac{\partial\mu}{\partial y}=\frac{r(1+q^2)-2pqs+t(1+p^2)}{(p^2+q^2+1)^{\frac{3}{2}}}$$

$$=\frac{1}{\rho_1}+\frac{1}{\rho_2},$$

where ρ_1, ρ_2 are the principal radii of curvature of the surface at (x, y, z).† Therefore

$$\varpi=T\left(\frac{1}{\rho_1}+\frac{1}{\rho_2}\right),$$

and resolution parallel to either of the other axes would have led to the same result.

102. *Application of the principle of energy to the case of a homogeneous liquid at rest in a vessel under the action of gravity.*‡

In equilibrium the value of the potential energy must be stationary, and it is composed of four parts: the gravitational energy $g\rho\iiint zdxdydz$, where z is the height of an element $dxdydz$; and the energy of the surfaces separating (α) liquid and air, (β) liquid and vessel, (γ) air and vessel.

Hence we require that

$$g\rho\iiint zdxdydz+AS_1+BS_2+CS_3$$

should be stationary, where S_1, S_2, S_3 denote the surfaces (α), (β), (γ)

* See any *Cours d'Analyse, e.g.* de la Vallée Poussin, t. i. p. 348 (4th ed.).

† See C. Smith, *Solid Geometry*, p. 225.

‡ This discussion of the theory of capillarity is taken from Mathieu, *Théorie de la Capillarité*, 1883.

and A, B, C their energies per unit area respectively; subject to the condition that the volume $\iiint dxdydz$ is constant.

For a slight displacement of the surface S_1, between the liquid and air, if δn denote the element of the normal to the surface S_1 between corresponding elements of S_1 in the old and new positions, the variation of the first term is clearly $g\rho\iint z\delta n dS_1$.*

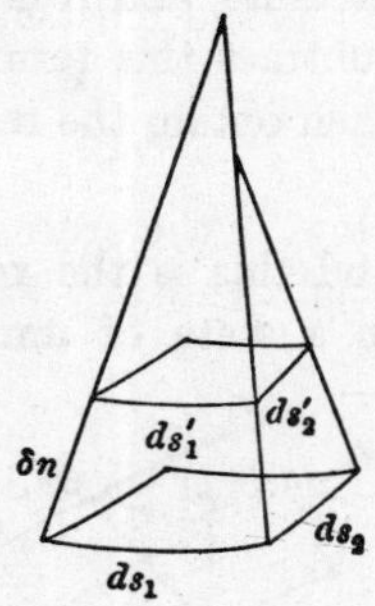

Suppose, in the first place, that the line of contact of the liquid with the vessel does not vary, then S_2 and S_3 are constant and S_1 changes to S_1'. Consider an element $ds_1 ds_2$ of S_1 bounded by lines of curvature; the normals through the boundaries of this element cut the surface S_1' in an element $ds_1'ds_2'$, and if ρ_1, ρ_2 are the principal radii of curvature,

$$ds_1'=\left(1-\frac{\delta n}{\rho_1}\right)ds_1, \qquad ds_2'=\left(1-\frac{\delta n}{\rho_2}\right)ds_2;$$

$$\therefore dS_1'-dS_1=ds_1'ds_2'-ds_1ds_2=-\left(\frac{1}{\rho_1}+\frac{1}{\rho_2}\right)\delta n ds_1 ds_2,$$

or
$$\delta dS_1=-\left(\frac{1}{\rho_1}+\frac{1}{\rho_2}\right)\delta n \,.\, dS_1.$$

But we require that

$$g\rho\iint z\delta n dS_1+A\delta\iint dS_1=0,$$

or, that
$$\iint\left\{g\rho z-A\left(\frac{1}{\rho_1}+\frac{1}{\rho_2}\right)\right\}\delta n dS_1=0, \qquad (1)$$

subject to the condition of constant volume, viz. $\iint \delta n dS_1=0$; and this is equivalent to

$$\iint\left\{g\rho(z-h)-A\left(\frac{1}{\rho_1}+\frac{1}{\rho_2}\right)\right\}\delta n dS_1=0,$$

where h is a constant and δn is arbitrary.

$$\therefore A\left(\frac{1}{\rho_1}+\frac{1}{\rho_2}\right)=g\rho(z-h);$$

* It is probable that the density of the liquid infinitely near the surface varies owing to the molecular action, but as the thickness of the layer of variable density is infinitely small compared with δn, we may neglect this variation without affecting the argument.

If we wish to take account of the atmospheric pressure Π above the liquid, we can do so by observing that, in the displacement considered, an amount of work $-\iint \Pi\delta n dS_1$ would be done by this pressure, and regarding this as a loss of potential energy we must subtract this term from the first member of equation (1), and we then obtain the result

$$A\left(\frac{1}{\rho_1}+\frac{1}{\rho_2}\right)=g\rho(z-h)+\Pi,$$

and this is the result that would hold good if the surface were in a state of uniform tension A and the pressure difference on opposite sides of the surface were $\Pi-g\rho(h-z)$.

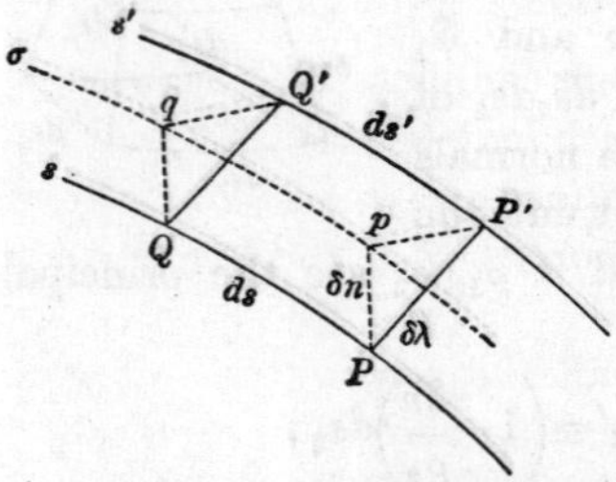

Secondly, suppose that the line of contact of the liquid with the vessel is displaced from s to s'. If we draw normals to the surface S_1 at all points of the line s, they will meet the surface S_1' in a line σ, and the surface S_1' may be considered as composed of two parts, the one Σ enclosed by the line σ, and the other Σ' between the lines σ and s'.* As before, we get

$$\Sigma-S_1=-\iint\left(\frac{1}{\rho_1}+\frac{1}{\rho_2}\right)\delta n dS_1;$$

and, if $\delta\lambda$ denote the distance between the elements ds, ds', Σ' may be considered as the projection of the elements $\delta\lambda ds$ of the surface of the vessel on the surface S_1', so that if i is the angle between the normals to the surfaces S_1 and S_2, then

$$\Sigma'=\int \cos i\delta\lambda ds.$$

Also $$\delta S_2=-\delta S_3=\int \delta\lambda ds.$$

Now since the potential energy is stationary we have

$$\delta\left\{g\rho\iiint zdxdydz+AS_1+BS_2+CS_3\right\}=0$$

subject to the condition that the mass is constant; or

* In the figure, PQ is an element ds of the line of contact s of the liquid with the vessel, and $P'Q'$, pq are corresponding elements of the lines s', σ respectively: $P'pqQ'$ is an element of the surface Σ'. The variation in the mass represented by the wedge-shaped elements $PP'q$ round the line of contact of the liquid and the vessel is of a higher order of small quantities than the rest and may be neglected.

$$g\rho\iint z\delta n dS_1 + A(\Sigma+\Sigma'-S_1)+B\delta S_2+C\delta S_3=0;$$

or $$\iint\left\{g\rho z - A\left(\frac{1}{\rho_1}+\frac{1}{\rho_2}\right)\right\}\delta n dS_1+\int(A\cos i+B-C)\delta\lambda ds=0$$

subject to the condition

$$\iint \delta n dS_1=0,$$

and, since $\delta\lambda$ is arbitrary, this gives equation (1) as before, and also

$$A\cos i+B-C=0 \quad . \quad . \quad . \quad (2)$$

or the angle between the surfaces of the liquid and the vessel is constant along the line of their intersection.

103. From the foregoing considerations combined with the results of experiment we are led to two laws which may be stated as follows:

(1) *At the bounding surface separating air from a liquid, or between two liquids, there is a surface tension which is the same at every point and in every direction.*

(2) *At the line of junction of the bounding surface of a gas and a liquid with a solid body, or of the bounding surface of two liquids with a solid body, the surface is inclined to the surface of the body at a definite angle, depending upon the nature of the solid and of the fluids.*

In the case of water in a glass vessel the angle is acute; in the case of mercury it is obtuse.

Assuming these laws we can account for many of the phenomena of capillarity and of liquid films.

104. *Rise of liquid between two plates.*

If t be the surface tension, a the constant angle at which the surface meets either plate, called the angle of capillarity, h the mean rise, and d the distance between the plates, we have, for the equilibrium of the unit breadth of the liquid,

$$2t\cos a=g\rho hd,$$

so that the rise increases with the diminution of the distance between the plates.

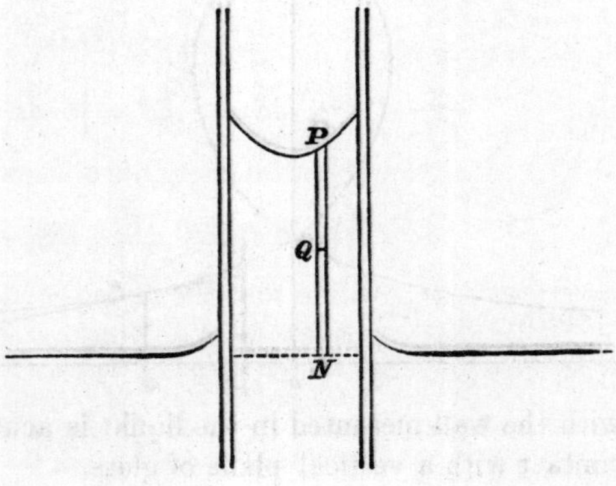

It will be seen that the pressure at any point Q is less than the pressure at N by $g\rho \cdot QN$,

$$\text{and } \therefore = \Pi - g\rho QN.$$

The atmospheric pressure at P being sensibly equal to the pressure at the water level outside, it follows that the weight PN is supported by the resultant of the surface tensions on its upper boundary.

105. *Rise of a liquid in a circular tube.*

In this case the column of liquid is supported by the tension round the periphery of its upper boundary, and therefore, if r be the internal radius,

$$2\pi rt \cos a = g\rho\pi r^2 h,$$

or

$$2t \cos a = g\rho rh.$$

The pressure at any point of the suspended column being less than the atmospheric pressure, it follows that if the column were high enough, the pressure would merge into a state of tension, which would still follow the law of fluid pressure of being the same in every direction.

It may be observed that the potential energy, due to the ascent of the column, is independent of the radius.

106. The Capillary Curve. The form of the surface assumed by a liquid in contact with a vertical wall can be investigated if we assume that the surface is cylindrical with horizontal generators. The cross-section of such a cylindrical surface is called the capillary curve.

We shall take the case in which the angle of contact of the liquid

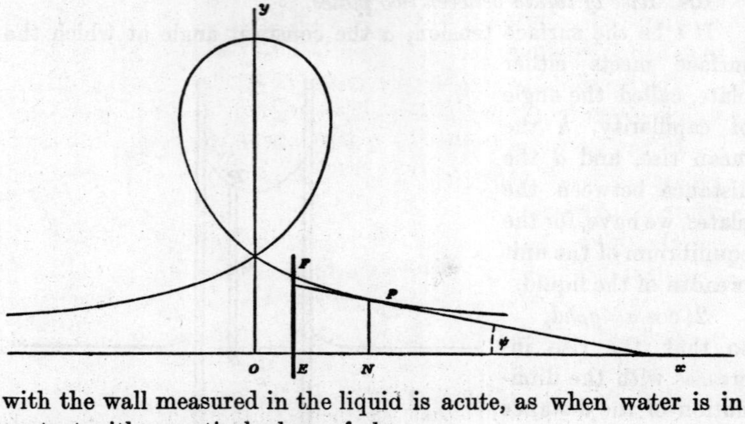

with the wall measured in the liquid is acute, as when water is in contact with a vertical plane of glass.

Let EF be the wall which the liquid meets at an angle α. Take a horizontal axis Ox at right angles to the wall at the natural level of the liquid, *i.e.* the level at which the pressure in the liquid is equal to the atmospheric pressure Π.

Let r be the radius of curvature at P (x, y) on the capillary curve, and let t be the surface tension. Then the theorem of Art. 101 gives

$$\frac{t}{r}=\Pi-p=g\rho y.$$

Hence, putting $4t=g\rho c^2$, we get

$$ry=\tfrac{1}{4}c^2 \quad . \quad . \quad . \quad . \quad . \quad (1)$$

If ψ denote the acute angle between the tangent at P and the axis of x as in the figure, and the arc s be measured from the wall, we have $r=-ds/d\psi$, and $ds/dy=-\operatorname{cosec}\psi$.

Therefore $$ydy=\tfrac{1}{4}c^2\sin\psi d\psi \quad . \quad . \quad . \quad . \quad (2)$$

Hence, since y and ψ vanish together

$$y^2=\tfrac{1}{2}c^2(1-\cos\psi)=c^2\sin^2\tfrac{1}{2}\psi,$$

therefore $$y=\pm c\sin\tfrac{1}{2}\psi \quad . \quad . \quad . \quad . \quad (3)$$

and in the case considered the upper sign must be taken. Again, $dy/dx=-\tan\psi$, so that

$$dx=-\tfrac{1}{2}c\cos\tfrac{1}{2}\psi\cot\psi d\psi$$
$$=-\tfrac{1}{4}c(\operatorname{cosec}\tfrac{1}{2}\psi-2\sin\tfrac{1}{2}\psi)d\psi.$$

Therefore $$x=\tfrac{1}{2}c\log\cot\tfrac{1}{4}\psi-c\cos\tfrac{1}{2}\psi \quad . \quad . \quad . \quad (4)$$

provided that the origin be chosen so that $x=0$ when $\psi=\pi$. The capillary curve is represented by equations (3) and (4). It has a loop as in the figure and is asymptotic to the axis of x.

The height above the natural level at which the liquid meets the wall is given by (3) in the form $EF=c\sin(\tfrac{1}{4}\pi-\tfrac{1}{2}\alpha)$.

In the case of a liquid, such as mercury, for which the angle of contact is obtuse, it is convenient to measure y downwards, and the figure is inverted.

The differential equation (1) is also the equation of equilibrium of a flexible rod bent by terminal forces. The integration in finite terms obtained above depends on our being able to assume that y and ψ vanish together. With any other constant of integration in the integral form of (2) we shall find that x is expressed by elliptic integrals: and the curve may assume a variety of forms, and is known as the *elastic curve* or the *elastica*.

107. Drop of Liquid. If a drop of liquid be placed on a horizontal plane, the equation of equilibrium will be

$$\frac{1}{\rho_1}+\frac{1}{\rho_2}=\frac{\varpi}{t},$$

where t is the surface tension, and ϖ is the difference between the internal pressure and the atmospheric pressure.

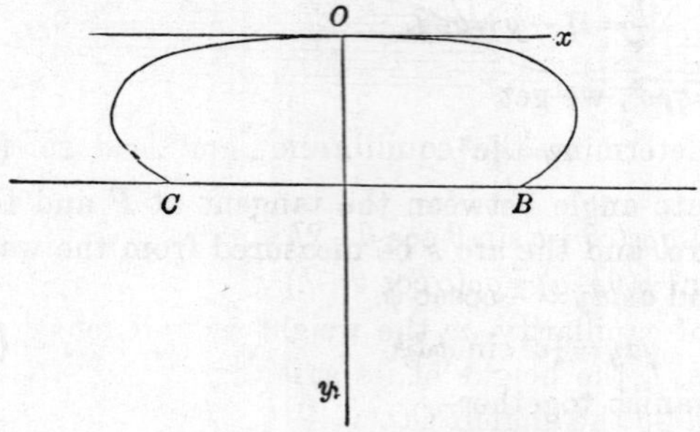

In general the drop will assume the form of a surface of revolution.

The only case, however, which is capable of simple treatment is that in which we may regard the drop as so large that it may be considered to have a flat top and that curvature in a horizontal sense is negligible. Thus, measuring y downwards from the top when the pressure is atmospheric, we have $\varpi=g\rho y$, and putting $4t=g\rho c^2$, we get as in Art. 106

$$ry=\tfrac{1}{4}c^2,$$

and the vertical section is the capillary curve.

With axes as in the figure we shall find that

$$y=c\sin\tfrac{1}{2}\psi,$$

and $$x=\tfrac{1}{2}c\log\tan\tfrac{1}{4}\psi+c\cos\tfrac{1}{2}\psi+\text{const.},$$

where ψ is the inclination of the tangent to Ox.

Thus, if α be the angle of contact of the liquid with the plane measured in the liquid, the height of the drop is $c\sin\frac{1}{2}\alpha$.

This would hold good for the case of mercury upon glass or water upon steel.

108. Floating needle. The well-known experiment of floating a needle on the surface of water can be explained by aid of the laws of surface tension.

The figure representing a section of the needle and the surface of the water at right angles to the axis of the needle, the forces in action on the needle are the tensions on P and Q, and the water pressure on PAQ, which is equal to the weight of the volume $NPAQM$ of water; these forces counterbalance the weight of the needle.

Further, the horizontal component of the tension at P, together with the horizontal water pressure on BD, is equal to the tension at B, PD being horizontal and BD vertical.

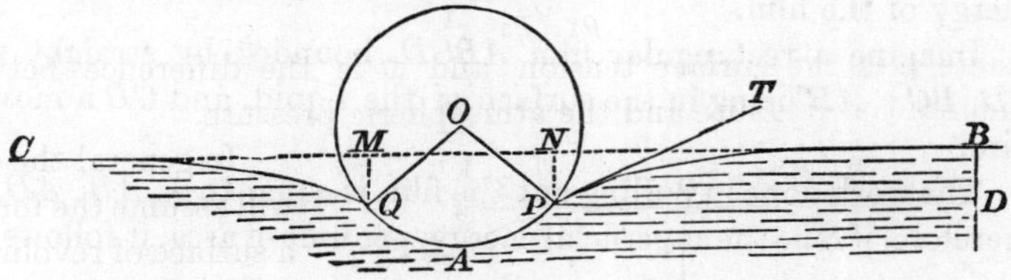

These conditions determine the equilibrium, and lead to the equations

$$2t \sin(\theta-\alpha)+g\rho c(c\theta+c\sin\theta\cos\theta-2h\sin\theta)=w,$$

$$4t\sin^2\tfrac{1}{2}(\theta-\alpha)=g\rho(c\cos\theta-h)^2,$$

where α is the angle of capillarity, w the weight of unit length of the needle c its radius, h the height of its axis above the natural level of the water, and 2θ the angle POQ.

109. Liquid films. Liquid films are produced in various ways; a soap bubble is a familiar instance, and liquid films may be formed, and their characteristics observed, by shaking a clear glass bottle containing some viscous fluid, or by dipping a wire frame into a solution of soap and water, or glycerine, and slowly drawing it out.

The fact that films apparently plane can be obtained, shows that the action of gravity may be neglected in comparison with the tension of the film.

It is found that a very small tangential action will tear the film, and it is therefore inferred that the stress across any line is entirely normal to that line. From this it follows that the tension is the same in every direction.

For if we consider a small triangular element of the surface, the equilibrium in the tangent plane is entirely determined by the tensions across the sides of the triangle, for the tangential impressed forces, if there be any, will ultimately vanish in comparison with the tensions; and since these tensions are at right angles to the sides, they must be in the ratio of their lengths, and therefore the measures of tension in all directions are the same.

Further, the tension will be the same at all points of the surface, for, if a small rectangular element be considered, the tensions on opposite sides must be equal.

110. Energy of a plane film. If a plane film be drawn out from a reservoir of viscous liquid, a certain amount of work is expended, and the work thus expended represents the potential energy of the film.

Imagine a rectangular film $ABCD$, bounded by straight wires AD, BC; AB being in the surface of the liquid, and CD a movable wire.

The work done in pulling out the film is equal to $\tau . AB . AD$, and therefore, if S be the superficial energy, per unit of area, it follows that

$$S=\tau.$$

It should be observed that what we have here called the tension of the film is equal to twice the surface tension of either side of the film.

111. Energy of a spherical soap-bubble. The energy of a soap-bubble is the work done in producing it. This consists of two parts, viz. the work done in pulling out the film and the work done in compressing the air in the bubble.

If t be the surface tension, the former part is tS, where S denotes the area of the surface, for the energy of a small plane element is $t\delta S$. For the latter part, let p denote the pressure of the air inside when the radius is r, and Π the atmospheric pressure, then $p-\Pi=\frac{2t}{r}$; and, if the bubble contains a mass of air which at pressure Π would occupy a volume V, then

$$\Pi V=\tfrac{4}{3}\pi r^3 p=pV', \text{ say,}$$

and by Art. 95 the work done in compressing the air from volume V to volume V'

$$=\Pi V \log\frac{V}{V'}-\Pi(V-V')$$

$$=\tfrac{4}{3}\pi r^3\left\{\left(\Pi+\frac{2t}{r}\right)\log\left(1+\frac{2t}{r\Pi}\right)-\frac{2t}{r}\right\}.$$

If we assume that the difference between the pressures inside and outside the bubble is small compared with the atmospheric pressure, we may take $\frac{2t}{r\Pi}$ as small, and the last expression becomes

$$\tfrac{4}{3}\pi r^3\left\{\left(\Pi+\frac{2t}{r}\right)\left(\frac{2t}{r\Pi}-\frac{2t^2}{r^2\Pi^2}\right)-\frac{2t}{r}\right\}$$

$$=\tfrac{4}{3}\pi r^3 \cdot \frac{2t^2}{r^2\Pi}=\tfrac{2}{3}\frac{t^2 S}{r\Pi},$$

so that the work done in compressing the air is to that done in pulling out the film as $2t : 3r\Pi$.

112. The forms of liquid films. Minimal surfaces. If the air pressure be the same on both sides of a film, the condition of equilibrium is that

$$\frac{1}{\rho_1}+\frac{1}{\rho_2}=0,$$

or that the mean curvature is zero.

This condition is satisfied in the cases of the catenoid and the helicoid, which are therefore possible forms of liquid films.

In Cartesian co-ordinates the equation becomes

$$\left\{1+\left(\frac{\partial z}{\partial y}\right)^2\right\}\frac{\partial^2 z}{\partial x^2}-2\frac{\partial z}{\partial x}\frac{\partial z}{\partial y}\frac{\partial^2 z}{\partial x\partial y}+\left\{1+\left(\frac{\partial z}{\partial x}\right)^2\right\}\frac{\partial^2 z}{\partial y^2}=0,$$

as in Art. 101.

The discussion of this equation is the subject of many memoirs by eminent mathematicians, and several very remarkable special solutions have been obtained.

For instance, the surfaces

$$e^z=\cos y \sec x \text{ and } \sin z=\sinh x \sinh y$$

will be each found to possess the property that its mean curvature is zero.

In Plateau's work, *Sur les liquides soumis aux seules forces moléculaires* (2 vols. 1873), will be found an elaborate account of the labours of mathematicians on this subject, and of his own extensive series of experiments ; and, in Darboux's *Théorie Générale des Surfaces*, tome i., livre iii., there is a full discussion of minimal surfaces, that is, of surfaces which satisfy the condition given above.

113. If the form of the film be that of a surface of revolution about the axis of x, and at any point (x, y) on the meridian curve the tangent makes an angle ψ with the axis of x, by resolving parallel to this axis for the equilibrium of a portion of the film between planes perpendicular to the axis, we get

$$2\pi y \, . \, t \cos \psi=\text{const.}$$

or

$$y=c \sec \psi.$$

Thus $$dy/ds \text{ or } \sin\psi = c\sec\psi\tan\psi d\psi/ds,$$

and $$\therefore s = c\tan\psi,$$

provided we measure s and ψ so that they vanish together. Hence the meridian curve must be a catenary; and a catenoid, or the surface obtained by revolving a catenary about its directrix, is the only possible form of revolution of a film when the pressure is the same on both sides.

In the case of a surface of revolution, one of the principal radii of curvature at a point is the normal intercepted between the point and the axis of revolution. It is easy to see that in the catenary the intercept on the normal between the curve and its directrix is equal in length to the radius of curvature, and the catenoid being an anticlastic surface the relation

$$\frac{1}{\rho_1}+\frac{1}{\rho_2}=0$$

is satisfied. We may also show conversely that this relation leads to the catenoid as the only solution.

114. The same result is obtained by the principle of energy, for the area

$$\int 2\pi y ds$$

is then a maximum or a minimum, and, by the Calculus of Variations, this leads to a catenary as the generating curve, the axis of revolution being the directrix of the catenary.

In Todhunter's *Researches in the Calculus of Variations* it is shown that it is not always possible, when a straight line and two points in the same plane are given, to draw a catenary which shall pass through the two points and have the straight line for its directrix.

It is also shown that, under certain conditions, two such catenaries can be drawn, and that, in a particular case, only one such catenary can be drawn. The two catenaries, when they exist, correspond to the figure formed by a uniform endless string hanging over two smooth pegs.

When there are two catenaries the surface generated by the revolution of the upper one about the directrix is a minimum, but the surface generated by the lower one is not a minimum. When there is only one catenary, it is not a minimum.

Hence it appears that if a framework be formed of two circular wires, the planes of which are parallel to each other and perpendicular to the line joining their centres, it is not always possible to connect the wires by a liquid film. In certain cases it is possible to connect the wires by one of two catenoids, but, in the case of the catenoid formed by the revolution of the upper catenary, the equilibrium is stable, while the other catenoid is unstable.

When there is only one catenoid it is unstable.

There is also a discontinuous solution of the problem, consisting of the two circles formed by the revolution of the ordinates of the points, and an infinitesimally slender cylinder connecting their centres.

In the article on Capillarity in the *Encyclopædia Britannica* * by Clerk Maxwell, the question is discussed in the following manner.

When two catenaries, having the same directrix, can be drawn through two given points, and the catenoids are formed by revolution about the directrix, the mean curvature of each catenoid is zero.

If another catenary be drawn between the two catenaries, passing through the same two points, its directrix will be above the directrix of the other two, and therefore its radius of curvature at any point will be less than the distance, along the normal, of the point from the first directrix.

The mean curvature of the surface of revolution is therefore convex to the axis, and it follows that if either catenoid is displaced into another catenoid between the two, the film will move away from the axis.

Again, if a catenoid be taken outside the two, its mean curvature will be concave to the axis, and therefore if the upper catenoid be displaced upwards and the lower one downwards the film will, in each case, move towards the axis.

Hence it follows that the outer of the two catenoids is stable, and that the inner one is unstable.

This argument, however, does not apply to any other form of displacement, and therefore, for a complete proof of the case of stability, it is necessary to have recourse to the methods of the Calculus of Variations.

* This article was revised by Lord Rayleigh in the eleventh edition of the *Encyclopædia*.

115. If the pressure on the two sides of a film be different, and if ϖ be the difference, the condition of equilibrium is

$$\frac{1}{\rho_1}+\frac{1}{\rho_2}=\frac{\varpi}{t},$$

or that the mean curvature is constant.

If the film be in the form of a surface of revolution, we can show that the meridian curve is the path of the focus of a conic rolling on a straight line.

Let S be the focus of the conic and P its point of contact with the given line.

Let $SP=r$ and let p be the perpendicular SY from S to the line. The (p, r) equation of a conic is of the form

$$\frac{l}{p^2}-\frac{2}{r}=\mp\frac{1}{a} \text{ or zero} \quad . \quad . \quad . \quad (1)$$

according as the conic is an ellipse, hyperbola, or parabola ; where l denotes the semi-latus rectum and a the semi-major axis.

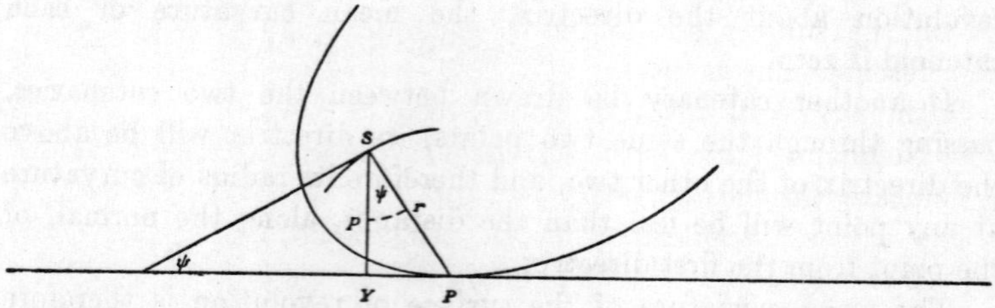

Hence, if ρ be the radius of curvature of the locus of S and ψ the angle YSP, we have

$$\frac{1}{\rho}=-\frac{d\psi}{ds}=-\frac{d\psi}{dp}\sin\psi=\frac{d}{dp}\left(\frac{p}{r}\right).$$

Also, if the locus of S is rotated round the fixed line, the normal SP is one of the principal radii of curvature of the surface of revolution, and

$$\therefore \frac{1}{\rho}+\frac{1}{r}=\frac{2}{r}-\frac{p}{r^2}\frac{dr}{dp}.$$

But from (1)
$$\frac{l}{p^3}=\frac{1}{r^2}\frac{dr}{dp},$$
whence we get

$$\frac{1}{\rho}+\frac{1}{r}=\pm\frac{1}{a} \text{ or zero},$$

according as the rolling curve is an ellipse, hyperbola, or parabola.

The third is the Catenoid ; the first and second are called by Plateau the Unduloid and the Nodoid, the former being a sinuous curve, and the latter presenting a succession of nodes.

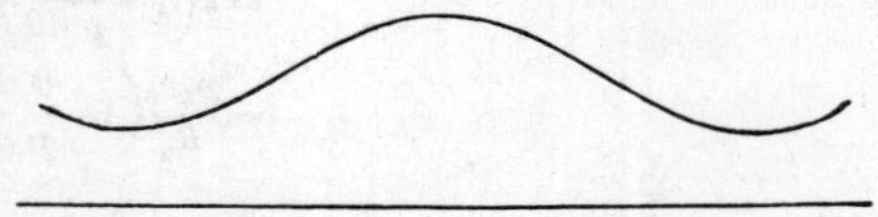

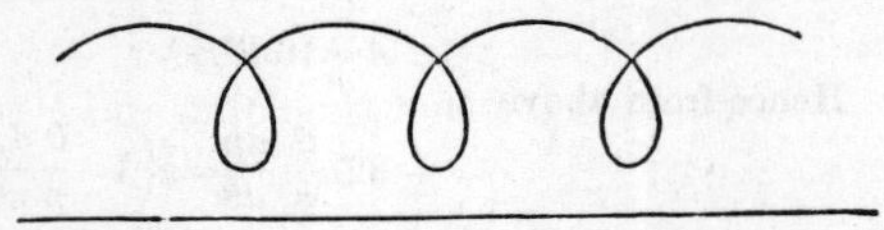

To obtain a clear view of the generation of the nodoid, it must be considered that, as one branch of the hyperbola rolls, the point of contact moves off to an infinite distance ; the line then becomes asymptotic to both branches, and the other branch begins to roll, thereby producing a perfect continuity of the figure.*

Of the numerous works and papers on the subject of liquid films the student will find full accounts in Plateau's work, and in Professor Clerk Maxwell's article in the *Encyclopædia Britannica* ; and on the subject of Capillarity generally the following works and references may be useful :

Mathieu, *Théorie de la Capillarité*, 1883.

F. Neumann, *Vorlesungen über die Theorie der Capillarität*, 1894.

Poincaré, *Capillarité*, 1895.

The articles *Kapillarität* by H. Minkowski in *Encyklop. der Math. Wissensch.*, Bd. v., 1907, and by F. Pockels in Winkelmann's *Handbuch der Physik*, Bd. i., 1908, both of which contain a full bibliography of the subject.

116. Example. *A soap bubble extends from fixed boundaries, so as with them to form a closed space whose volume is v_0, and contains a gas at pressure p_0 and absolute temperature θ_0. The temperature of the gas is gradually raised. If A be the area of the film when the temperature is θ, and pressure p, show that*

$$t\theta_0\frac{dA}{d\theta}=p_0v_0\left(1-\frac{\theta}{p}\frac{dp}{d\theta}\right),$$

where t is the surface tension supposed constant, and the external pressure is neglected. Deduce the relation between p and θ when the bubble is spherical.

The change of energy

$$=t\delta A$$

$$=p\delta v.$$

But

$$pv=k\theta\,;$$

$$\therefore\ p\delta v=k\delta\theta-v\delta p\,;$$

* Plateau, vol. i. p. 136. See also an article by Delaunay, *Liouville's Journal*, 1841, and an article by Lamarle, *Bulletins de l'Académie Belgique*, 1857.

$$\therefore t\frac{dA}{d\theta}=k-v\frac{dp}{d\theta}$$

$$=k\left(1-\frac{\theta}{p}\frac{dp}{d\theta}\right)$$

$$=\frac{p_0v_0}{\theta_0}\left(1-\frac{\theta}{p}\frac{dp}{d\theta}\right).$$

For a sphere $A=4\pi r^2$, and $p=\frac{2t}{r}$;

$$\therefore A=16\pi t^2/p^2.$$

Hence from above

$$-32\pi\frac{t^3}{p^3}\frac{dp}{d\theta}=k\left(1-\frac{\theta}{p}\frac{dp}{d\theta}\right);$$

$$\therefore -2\frac{At}{p}\frac{dp}{d\theta}=k\left(1-\frac{\theta}{p}\frac{dp}{d\theta}\right),$$

but

$$pv=k\theta;$$

$$\therefore \tfrac{1}{3}prA=k\theta \text{ or } \tfrac{2}{3}tA=k\theta;$$

$$\therefore -\frac{3k\theta}{p}\frac{dp}{d\theta}=k-\frac{k\theta}{p}\frac{dp}{d\theta};$$

$$\therefore \frac{2\theta}{p}\frac{dp}{d\theta}+1=0;$$

$$\therefore p^2\theta=\text{constant}.$$

EXAMPLES

1. Two spherical soap-bubbles are blown, one from water, and the other from a mixture of water and alcohol: if the tensions per linear inch are equal to the weights of one grain and $\frac{7}{12}$ grain respectively, and if the radii be $\frac{7}{9}$ inch and $1\frac{1}{3}$ inch respectively, compare the excess, in the two cases, of the total internal over the total external pressure.

2. If two soap-bubbles of radii r and r', are blown from the same liquid, and if the two coalesce into a single bubble of radius R, prove that, if Π be the atmospheric pressure, the tension is equal to

$$\frac{\Pi}{2}\cdot\frac{R^3-r^3-r'^3}{r^2+r'^2-R^2}.$$

3. The superficial tensions of the surfaces separating water and air being 8·25, water and mercury 42·6, mercury and air 55, what will be the effect of placing a drop of water upon a surface of mercury?

4. Show that if a light thread with its ends tied together form part of the internal boundary of a liquid film, the curvature of the thread at every point will be constant.

If the thread have weight, and if the film be a surface of revolution about a vertical axis, prove that, in the position of equilibrium, the tension of the thread is

$$\frac{l}{2\pi}\sqrt{\tau^2-w^2},$$

l being its length, w its weight per unit length, and τ the tension of the film.

5. A plane liquid film is drawn out from a soap-sud reservoir ; prove that the numerical value of the energy per unit of area (e) is equal to that of the tension (T) per unit of length.

If the film be removed from the reservoir, and if σ denote subsequently the mass of unit of area, prove that

$$T=e-\sigma\frac{de}{d\sigma}. \quad \text{(Clerk Maxwell.)}$$

6. Any number of soap-bubbles are blown from the same liquid and then allowed to combine with one another. Find an equation for determining the radius of the resulting bubble, and prove that the decrease of surface bears a constant ratio to the increase of volume.

7. If a film under unequal internal and external pressure form a surface of revolution, prove that the inclination ϕ of the tangent plane at P to the axis is given by the equation

$$\cos\phi=\frac{x}{a}+\frac{b}{x}:$$

x being the perpendicular from P on the axis and a, b constants.

8. Two soap-bubbles are in contact ; if r_1, r_2 be the radii of the outer surfaces, and r the radius of the circle in which the three surfaces intersect,

$$\frac{3}{4r^2}=\frac{1}{r_1^{\,2}}+\frac{1}{r_2^{\,2}}-\frac{1}{r_1 r_2}.$$

9. If water be introduced between two parallel plates of glass, at a very small distance d from each other, prove that the plates are pulled together with a force equal to

$$\frac{2At\cos\alpha}{d}+Bt\sin\alpha,$$

A being the area of the film and B its periphery.

10. A needle floats on water with its axis in the natural level of the surface ; if σ be the specific gravity of steel referred to water, β the angle of capillarity, and $2a$ the angle subtended at the axis by the arc of a cross-section in contact with the water, prove that

$$(\pi\sigma-a)\sin\tfrac{1}{2}(a-\beta)=\cos a\cos\tfrac{1}{2}(a+\beta).$$

11. A soap-bubble is filled with a mass m of a gas whose pressure is $k\times$(its density) at the temperature considered. The radius of the bubble is a, when it is first placed in air. The barometer then rises, the temperature remaining unaltered. Show that the radius of the bubble increases or diminishes according as the tension of the film is greater or less than $\dfrac{9}{8}\dfrac{km}{\pi a^2}$.

12. Prove that the equation

$$y=x\tan(az+b)$$

represents a possible form of a liquid film, the pressure on both sides being the same.

13. If two needles floating on water be placed symmetrically parallel to each other, show that they will be apparently attracted to each other, and that this is due to the surface tension.

14. A small cube floats with its upper face horizontal, in a liquid such that its angle of contact with the surface of the cube is obtuse and equal to $\pi-a$.

If ϱ is the density of the liquid, and σ that of the cube, and if $g\varrho c^2$ is the surface tension, prove that the cube will float if

$$\frac{\sigma}{\varrho}<1+4\frac{c^2}{a^2}\cos\alpha+2\frac{c}{a}\sin\left(\frac{\pi}{4}-\frac{\alpha}{2}\right).$$

15. Two equal circular discs of radius a are placed with their planes perpendicular to the line which joins their centres, and their edges are connected by a soap-film which encloses a mass of air that would be just sufficient in the same atmosphere to fill a spherical soap-bubble of radius c. If the film be cylindrical when the distance between the discs is b, prove that in order that it may become spherical the distance between the discs must be lessened to $2z$, where

$$z(3a^2+2z^2)\left\{8c^2-3ab+\frac{6a^2b-8c^3}{\sqrt{a^2+z^2}}\right\}=6abc^2(2a-c).$$

16. A liquid film of total surface tension T is in the form of a cylinder joining two equal parallel circular discs of radius $2a$, with their centres at a distance $2a$ apart on a line perpendicular to their planes. A pin-hole is made in one of the discs so that the air slowly escapes; show that a total quantity

$$\varrho_0\pi a[8a^2\{1+T/(2a\Pi)\}-c^2\{1+2\sinh(a/c)\}]$$

will escape, where ϱ_0 and Π are atmospheric density and pressure, and c is given by $\cosh(a/c)=2a/c$.

17. A plane plate is partly immersed in a liquid of density ϱ and surface tension t. The angle of capillarity for the liquid and substance of the plate is β, and the plate is inclined at an angle α to the horizontal. Prove that the difference of the heights of the liquid on the two sides of the plate above the undisturbed surface level is

$$4\left\{\frac{t}{g\varrho}\right\}^{\frac{1}{2}}\cos\frac{\pi-2\beta}{4}\sin\frac{\pi-2\alpha}{4}.$$

18. A volume $\frac{4}{3}\pi c^3$ of gravitating liquid of astronomical density ϱ is surrounded by an atmosphere at pressure Π and contains a concentric cavity filled with air, whose volume at this atmospheric pressure is $\frac{4}{3}\pi a^3$. The surface tension of the liquid is t. Prove that the radius x of the cavity in the configuration of equilibrium is given by the equation

$$\Pi\left(\frac{a^3}{x^3}-1\right)=2t\left\{\frac{1}{x}+\frac{1}{\sqrt[3]{(c^3+x^3)}}\right\}+\frac{2}{3}\pi\varrho^2\left\{\frac{c^3+3x^3}{\sqrt[3]{(c^3+x^3)}}-3x^2\right\}.$$

19. A liquid film hangs in the form of a surface of revolution with its axis vertical. The upper boundary of the film is a circular wire held horizontally, the lower boundary is a heavy elastic thread, hanging freely in the form of a horizontal circle of radius r. The natural length of the thread is $2\pi a$, its modulus of elasticity is λ, and its weight is $2\pi aw$. The tension of the film is t. Prove that r satisfies the equation

$$(\lambda^2-a^2t^2)r^2-2\lambda^2ar+(\lambda^2+w^2a^2)a^2=0.$$

20. A wire circle (radius a) is placed in the surface of soapy water and raised gently, so as to draw after it a film. Prove that, neglecting its weight, the meridian section of the film is a catenary, and investigate the angle at which the film meets the undisturbed surface of the water. Also prove that the parameter of the meridian catenary, when the area of the film is equal to πa^2, is a/z, where z is given by

$$\cosh^{-1}z+z(z^2-1)^{\frac{1}{2}}=z^2.$$

21. Two circular rings with a common axis at right angles to their planes support a closed liquid film containing air at a greater pressure than the external air: show that the ends of the film are spheres of radius $a=\frac{2T}{p}$, and that the surface between the rings is a surface of revolution of which the meridian curve has an intrinsic equation $\sin\phi=\frac{x}{a}\pm\frac{b}{x}$, where ϕ is the inclination of the normal to the axis and x is the distance from the axis.

22. A long circular cylinder of radius r entirely immersed in liquid, whose acute angle of contact with it is α, is gradually made to emerge, its axis being kept horizontal. Show that contact with the liquid finally ceases when the axis reaches a height h above the original and ultimate level of the liquid given by the equations

$$h=r\cos(\phi-\alpha)+c\cos\frac{\phi}{2},$$

$$\frac{2r}{c}\sin(\phi-\alpha)+2\sin\frac{\phi}{2}-\tanh^{-1}\sin\frac{\phi}{2}=2\sin\frac{\pi}{4}-\tanh^{-1}\sin\frac{\pi}{4},$$

the ratio of the surface tension to the density of the liquid being $\frac{1}{4}gc^2$.

23. A long wedge of vertical angle 2α, floats in water with its base horizontal and its top edge in the natural level of the surface. Prove that, if the capillary action at the ends be neglected,

$$w-w'=2T\sec\alpha(\sin\alpha+\cos\gamma),$$

where w is the weight of the wedge per unit length, w' that of an equal volume of water, T the surface tension, and γ the supplement of the angle of capillarity.

24. A drop of fluid under no forces except uniform external pressure and surface tension rotates as a rigid body about an axis; show that on the surface $3/R_2-1/R_1$ is constant, where R_1, R_2 are the principal radii of curvature of the surface.

25. Prove that, when the axis of z is along a downward vertical, and the origin suitably chosen, the surface of separation of two fluids of densities μ_1, μ_2 satisfies the relation

$$2z=a^2(\varrho^{-1}+\varrho'^{-1}),$$

where ϱ, ϱ' are the principal radii of curvature taken positive when the concavity is downwards, $a^2=2T/\{g(\mu_1-\mu_2)\}$, and T is the capillary constant of the interface.

If the surface is one of revolution about the z axis, show that the approximate equation (in cylindrical co-ordinates) of the part near the axis is of the form

$$2(z-z_0)=z_0a^{-2}r^2+\tfrac{1}{8}(z_0a^2+2z_0^3)a^{-6}r^4,$$

and indicate how, in the case of liquid in a tube, z_0 can be expressed in terms of the angle of contact.

CHAPTER VIII

THE EQUILIBRIUM OF REVOLVING LIQUID, THE PARTICLES OF WHICH ARE MUTUALLY ATTRACTIVE

117. If a liquid mass, the particles of which attract each other according to a definite law, revolve uniformly about a fixed axis, it is conceivable that, for a certain form of the free surface, the liquid particles may be in a state of relative equilibrium; since, however, the resultant attraction of the mass upon any particle depends in general upon its form, which is unknown, a complete solution of the problem cannot be obtained.

For any arbitrarily assigned law of attraction, the question is one of purely abstract interest, and it is only when the law is that of gravitation that it becomes of importance, from its relation to one of the problems of physical astronomy.

We shall consider the fluid homogeneous, and confine our attention to two cases; in the first of these the attractive forces are supposed to vary directly as the distance, and, in the second, to follow the Newtonian law.

118. *A homogeneous liquid mass, the particles of which attract each other with a force varying directly as the distance, rotates uniformly about an axis through its centre of mass; required to determine the form of the free surface.*

The resultant attraction on any particle is in the direction of, and proportional to, the distance of the particle from the centre of mass; and if μ be a measure of the whole mass of fluid, μx, μy, μz may represent the components of the attraction, parallel to the axis, on a particle of fluid about the point x, y, z.

Taking the origin at the centre of gravity, and axis of rotation as the axis of z, the equation of equilibrium is

$$dp=\rho\{(\omega^2x-\mu x)dx+(\omega^2y-\mu y)dy-\mu z dz\};$$

and therefore

$$p=C+\tfrac{1}{2}\rho\{(\omega^2-\mu)(x^2+y^2)-\mu z^2\}.$$

At the free surface p is zero or constant, and the equation to the free surface is

$$\left(1-\frac{\omega^2}{\mu}\right)(x^2+y^2)+z^2=D,$$

the constant D depending upon ω, and upon the mass of the fluid.

When ω is very small, the free surface is nearly spherical, and as ω^2 increases from 0 to μ, the spheroidal surface becomes more oblate.

When $\omega^2=\mu$, the free surface consists of two planes ; to render this possible we may conceive the fluid enclosed within a cylindrical surface, the axis of which coincides with the axis of rotation.

When $\omega^2>\mu$, the free surface is a hyperboloid of two sheets, which for a certain value (ω') of ω becomes a cone, the fluid filling the space between the cone and the cylinder. Taking account of the volume of the fluid, the value of ω' can be determined by putting $D=0$, since the pressure in this case vanishes at the origin.

If $\omega>\omega'$, the surface is a hyperboloid of one sheet, which, as ω increases, approximates to the form of a cylinder, and it is therefore necessary, for large values of ω, to conceive the containing cylinder closed at its ends.

The results of this article, it may be observed, are equally true of heterogeneous fluid, whatever be the law of variation of density in the successive strata.

119. *A mass of homogeneous liquid, the particles of which attract each other according to the Newtonian law, rotates uniformly, in a state of relative equilibrium, about an axis through its centre of mass ; required to determine a possible form of the surface.*

For the reason previously mentioned a direct solution of this problem cannot be obtained, but it can be shown that an oblate spheroid is a possible form of equilibrium.

Let the equation to the spheroid be

$$\frac{z^2}{c^2}+\frac{x^2+y^2}{c^2(1+\lambda^2)}=1,$$

the axis of rotation being the axis of z.

Then the resultant attractions, towards the origin, on a particle at the point (x, y, z) will be represented by

$$X=\frac{2\pi\rho x}{\lambda^3}\{(1+\lambda^2)\tan^{-1}\lambda-\lambda\},$$

$$Y=\frac{2\pi\rho y}{\lambda^3}\{(1+\lambda^2)\tan^{-1}\lambda-\lambda\},$$

$$Z=\frac{4\pi\rho z}{\lambda^3}\{\lambda-\tan^{-1}\lambda\}(1+\lambda^2),$$

parallel, respectively, to the axes.*

The equation of equilibrium is

$$dp=\rho\{(\omega^2x-X)dx+(\omega^2y-Y)dy-Zdz\}.$$

But from the equation to the spheroid,

$$xdx+ydy+(1+\lambda^2)zdz=0,$$

and as this must be a surface of equipressure, we must have

$$\omega^2-X/x=\omega^2-Y/y=-Z/(1+\lambda^2)z.$$

Hence we get

$$\frac{\omega^2}{2\pi\rho}=\frac{(1+\lambda^2)\tan^{-1}\lambda-\lambda}{\lambda^3}-\frac{2(\lambda-\tan^{-1}\lambda)}{\lambda^3},$$

or
$$\frac{\omega^2}{2\pi\rho}=\frac{(3+\lambda^2)\tan^{-1}\lambda-3\lambda}{\lambda^3} \quad . \quad . \quad . \quad (\alpha)$$

If ω and ρ are given, this equation determines λ and thence the ratio of the semiaxes of the spheroid is known.

To investigate the real solutions, let

$$y=\frac{(3+x^2)\tan^{-1}x-3x}{x^3} \quad . \quad . \quad . \quad (\beta)$$

Substituting the series for $\tan^{-1}x$, which is known to be convergent when $x<1$, we get

$$y=\sum_1^\infty(-)^{n-1}\frac{4n}{(2n+1)(2n+3)}x^{2n} \quad . \quad . \quad . \quad (\gamma)$$

Also
$$\frac{dy}{dx}=\frac{(7x^2+9)}{x^3(x^2+1)}-\frac{(x^2+9)}{x^4}\tan^{-1}x$$
$$=\frac{x^2+9}{x^4}\left\{\frac{7x^3+9x}{(x^2+1)(x^2+9)}-\tan^{-1}x\right\} \quad . \quad . \quad (\delta)$$
$$=\frac{x^2+9}{x^4}f(x),$$

where
$$f(x)=\frac{7x^3+9x}{(x^2+1)(x^2+9)}-\tan^{-1}x.$$

The forms (γ) and (β) show that y vanishes for $x=0$, and $x=\infty$,

* These expressions will be found in Laplace's *Mécanique Céleste*, Poisson's *Mécanique*, Duhamel's *Mécanique*, and Todhunter's *Statics*. In the last named, the equation to the spheroid is $(x^2+y^2)a^2+z^2/a^2(1-e^2)=1$, but the expressions used in the text will result from the expressions there given by putting

$$1-e^2=1/(1+\lambda^2).$$

By the use of λ, irrational quantities are avoided. Equivalent forms are given in Kelvin and Tait's *Natural Philosophy*, § 527, and Routh's *Analytical Statics*, vol. ii. § 219.

respectively; we shall show that as x increases from zero y has one maximum value and only one.

The sign of $\frac{dy}{dx}$ depends only on that of $f(x)$,

also when $\qquad x=0, \quad f(x)=0,$

and when $\qquad x=\infty, f(x)=-\frac{\pi}{2}.$

Again, we find that

$$f'(x)=\frac{8x^4(3-x^2)}{(x^2+1)^2(x^2+9)^2},$$

and this is positive from $x=0$ to $x=\sqrt{3}$, and negative for all greater values of x, so that $f(x)$ begins by being positive and increases as x increases to $\sqrt{3}$ and then decreases continuously; $f(x)$ therefore vanishes for a value of x greater than $\sqrt{3}$. By the help of tables we can easily show that $f(2)$ is positive and $f(3)$ negative, so that the value lies between 2 and 3. Also $f(2{\cdot}5)={\cdot}0025$ approximately, and Newton's method of approximation gives for the root

$$2{\cdot}5-\frac{f(2{\cdot}5)}{f'(2{\cdot}5)}=2{\cdot}5+{\cdot}0293=2{\cdot}5293\ldots$$

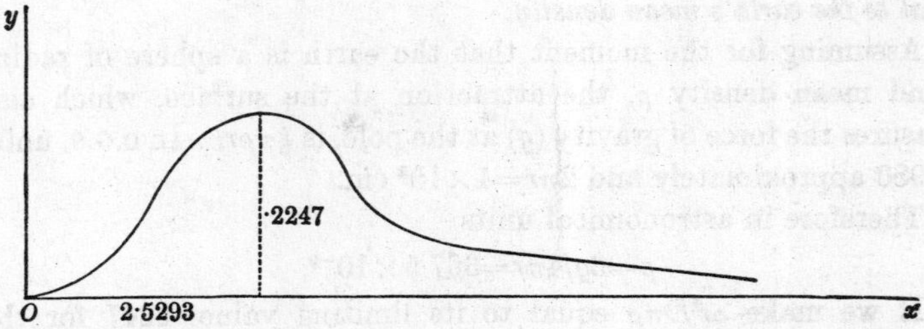

Hence $\frac{dy}{dx}$ vanishes only when $x=2{\cdot}5293\ldots$ and y is then a maximum and its value is ·2247.

The graph of equation (β) is therefore as in the figure, in which however the ordinate is drawn on a larger scale than the abscissa.

We conclude that if $\omega^2/2\pi\rho>{\cdot}2247$ the oblate spheroid is not a possible form of equilibrium, but if $\omega^2/2\pi\rho<{\cdot}2247$ there are two spheroidal forms possible, for there are two real values λ_1, λ_2 of the abscissa corresponding to every value of the ordinate less than ·2247.

120. The ellipticity of the spheroidal forms. When there are two real values λ_1, λ_2 of λ, one is greater and the other less than 2·5293. Let λ_2 be $>\lambda_1$, then as $\omega^2/2\pi\rho$ is diminished we see from the graph that λ_1 decreases and λ_2 increases, and since $\lambda_2 > 2{\cdot}5293$ therefore $\sqrt{1+\lambda_2^2} > 2{\cdot}72$; but the ratio of the semiaxes is $\sqrt{1+\lambda^2} : 1$, so that the larger value of λ always represents a much flattened spheroid, and the smaller we take $\omega^2/2\pi\rho$ the flatter does the spheroid become that corresponds to the root λ_2. On the other hand, for small values of $\omega^2/2\pi\rho$ the root λ_1 will be small, and if ϵ denote the ellipticity of the spheroid, we have

$$c(1+\epsilon) = c\sqrt{1+\lambda_1^2}, \text{ so that } \epsilon = \tfrac{1}{2}\lambda_1^2 \text{ approximately,}$$

and therefore from (γ)

$$\omega^2/2\pi\rho = \sum_1^\infty (-)^{n-1} \frac{4n}{(2n+1)(2n+3)} \lambda_1^{2n} = \frac{8\epsilon}{15},$$

as far as the first power of ϵ; or

$$\epsilon = 15\omega^2/16\pi\rho \text{ approximately.*}$$

Maclaurin was the first to prove that an oblate spheroid is a possible form of equilibrium of a rotating mass of homogeneous fluid, and the spheroids are therefore commonly called **Maclaurin's Spheroids.**

121. *Application to the case of a fluid, the density of which is equal to the earth's mean density.*

Assuming for the moment that the earth is a sphere of radius r and mean density ρ, the attraction at the surface, which also measures the force of gravity (g) at the pole, is $\frac{4}{3}\pi\rho r$. In C.G.S. units $g = 980$ approximately and $2\pi r = 4 \times 10^9$ cm.

Therefore in astronomical units

$$\rho = 3g/4\pi r = 367{\cdot}5 \times 10^{-9}.$$

If we make $\omega^2/2\pi\rho$ equal to its limiting value $\cdot 2247$ for the spheroidal form, and use the value just found for ρ, we obtain for the time of rotation $2\pi/\omega = 2$ hrs. 25 mins. This is therefore the smallest time in which a homogeneous mass, of density equal to the earth's mean density, could rotate uniformly in the form of an oblate spheroid.

Again, if we take for ω the earth's angular velocity $\dfrac{2\pi}{24 \times 60^2}$, we obtain

$$\frac{\omega^2}{2\pi\rho} = \frac{2\pi \times 10^9}{24^2 \times 60^4 \times 367{\cdot}5} = {\cdot}0023 \text{ approximately,}$$

which is less than the critical value $\cdot 2247$, so that for this density

* For a discussion in which the value of $\omega^2/2\pi\rho$ is obtained correct to the third power of the ellipticity, see Darwin's *Scientific Papers*, vol. iii. p. 423.

and angular velocity two spheroidal forms are possible, there being two real values for λ as explained in Art. 120. The larger value corresponds to a very flat spheroid, and the smaller gives a spheroid whose ellipticity is by Art. 120

$$\frac{15\omega^2}{16\pi\rho}=\frac{15}{8}\times\cdot 0023=\cdot 0043 \text{ or } \frac{1}{232} \text{ nearly.}$$

The earth, as is known by geodetic measurements, differs very slightly in its form from a sphere, its ellipticity being $\frac{1}{299\cdot 15}$,* that is, the axes of the spheroid are in the ratio 300·15 : 299·15. The fact that the axes of the homogeneous fluid spheroid, of the same mean density as the earth and rotating in the same time, are, as we have just seen, in the ratio 233 : 232 shows that it is extremely unlikely that the earth was at any period of its history a homogeneous fluid mass.

122. The prolate spheroid not a possible form. It must be observed that we have not solved the general problem of the form of a mass of rotating fluid in relative equilibrium, but merely shown that if $\omega^2/2\pi\rho<\cdot 2247$ an oblate spheroid is a possible form. And we notice that this result is independent of the mass of the fluid and depends only on the density and angular velocity. If $\omega^2/2\pi\rho>\cdot 2247$, it does not follow that equilibrium is impossible but only that there is no oblate spheroidal form possible in this case.

To examine whether a prolate spheroid is a possible form we may write $-\lambda'^2$ instead of λ^2 in Art. 119, where λ' is to be <1. Equations (α) and (γ) of that Article then give

$$\frac{\omega^2}{2\pi\rho}=-\sum_1^\infty\frac{4n}{(2n+1)(2n+3)}\lambda'^{2n},$$

which is impossible because the opposite sides of the equation are of unlike signs. Hence a prolate spheroid is not a possible form of equilibrium.

123. An important distinction has been pointed out by Poisson (tome ii. p. 547) between the surfaces of equal pressure in a fluid at rest under the action of extraneous forces, and in a fluid at rest, or revolving uniformly about a fixed axis, under the action of the mutually attractive forces of its particles.

Let *ABC* be the free surface, and *DEF* any surface of equal

* See *Encyc. Brit.* article, "Figure of the Earth," by A. R. Clarke and F. R. Helmert.

pressure; then, in the former case, the resultant force at any point of *DEF* is perpendicular to the surface at that point, and is unaffected by the existence of the fluid between *ABC* and *DEF*; this fluid could therefore be removed without affecting the equilibrium of the fluid mass bounded by *DEF*. In the latter case, the force at any point of *DEF*, although perpendicular to the surface at that point, is the resultant of the attractions of the mass of fluid contained by *DEF*, and of the mass contained between *DEF* and *ABC*; these two components of the resultant force are not necessarily perpendicular to the surface, and the fluid external to *DEF* cannot in general be removed without affecting the equilibrium of the remainder.

If, however, the fluid be homogeneous, and the particles attract each other according to the Newtonian law, so that the free surface may be spheroidal, the surfaces of equal pressure will be similar spheroids; and in this case, since the resultant attraction of an ellipsoidal shell, bounded by two concentric, similar, and similarly situated ellipsoids, on an internal particle is zero, the portion of fluid between *ABC* and *DEF* may be removed, provided the rate of rotation remain unaltered.

Moreover we have shown, Art. 120, that for a given value of ω not exceeding a determined limit, there are two possible spheroidal forms: let *ABC*, the free surface, have one of these forms, and describe within the fluid mass a concentric spheroid, *GHK*, similar to the other spheroid; then the fluid between *ABC* and *GHK* may be removed without affecting the fluid mass *GHK*.

The action of the shell upon a particle at a point *P* of the surface *GHK* is not perpendicular to the surface at *P*, but this action, combined with the attraction of the mass *GHK*, and the hypothetical force measured by $\omega^2 r$, is perpendicular to the surface, at *P*, of the spheroid passing through *P*, which is concentric with, and similar to, the surface *ABC*.

In other words, the direction of sensible gravity, that is, of the weight, of a particle on the surface is normal to the surface, and of a particle inside, normal to the surface of equal pressure which passes through the particle.

In the same manner if the free surface, *ABC*, have one of the possible forms, we can imagine a concentric shell of liquid added to the mass, and having its outer surface of the same form, or of the other possible form.

In the former case, ABC will still be a surface of equal pressure, but, in the latter case, ABC will cease to be a surface of equal pressure, since the new surfaces of equal pressure will be similar and similarly situated to the outer surface.

124. If a fluid mass be set in motion, about an axis through its centre of mass, with an angular velocity such as to make the value of $\omega^2/2\pi\rho$ greater than the limit obtained in Art. 119, it does not follow that the fluid cannot be in equilibrium in the form of a spheroid, for it may be conceived that the mass will expand laterally with reference to the axis, taking a more flattened shape, until its angular velocity is so far diminished as to render the spheroidal form possible.

If the mass consist of perfect fluid, its form will oscillate through the spheroid of equilibrium, but if, as is the case in all known fluids, friction be called into play by the relative displacement of the particles, the oscillations will gradually diminish and at length a position of equilibrium will be attained. Employing the principle that the angular momentum of the system, relative to the axis, will remain constant, we can determine the final angular velocity, and the form ultimately assumed.

Considering the question generally, suppose the mass of fluid set in motion in any way, and then left to itself; the centre of mass will be either at rest or moving uniformly in a straight line, and all we have to consider is the motion relative to the centre of mass.

Draw through the centre of mass the plane, in the direction of which the angular momentum is a maximum; then, however during the subsequent motion the fluid particles act on each other, this plane, which may be called the "momental" plane, will remain fixed, and when the motion of the particles relative to each other has been destroyed by their mutual friction, the axis perpendicular to this plane will be the axis of rotation of the fluid mass in its state of relative equilibrium.

Let H be the given angular momentum of the system, and ω its ultimate angular velocity.

Taking c and $c\sqrt{(1+\lambda^2)}$ for the axes of the spheroid of equilibrium, and M for the mass, the expression for the angular momentum is $\frac{2}{5}Mc^2(1+\lambda^2)\omega$;

$$\therefore \tfrac{2}{5}Mc^2(1+\lambda^2)\omega = H;$$

we have also
$$\tfrac{4}{3}\pi\rho c^3(1+\lambda^2) = M,$$

and from these two equations, combined with the equation

$$\frac{\omega^2}{2\pi\rho}=\frac{(3+\lambda^2)\tan^{-1}\lambda-3\lambda}{\lambda^3}\ \ldots \text{ Art. 119,}$$

the values of c, ω, and λ can be determined.

From the first two we obtain

$$\frac{\omega^2}{2\pi\rho}=\frac{25H^2(\frac{4}{3}\pi\rho)^{\frac{2}{3}}}{6M^{\frac{10}{3}}}(1+\lambda^2)^{-\frac{2}{3}};$$

$$\therefore \left\{\frac{(3+\lambda^2)\tan^{-1}\lambda-3\lambda}{\lambda^3}\right\}(1+\lambda^2)^{\frac{2}{3}}=\frac{25H^2}{6M^3}\left(\frac{4\pi\rho}{3M}\right)^{\frac{2}{3}}$$

is the equation which determines λ.

The equation always has a root, for the left-hand member vanishes and becomes infinite with λ, so that it ought to take a value equal to the positive constant on the right-hand side for some value of λ between zero and ∞. It can be shown, moreover, that there is only one positive root, for the derivative of the left-hand member can be shown to be positive always. Therefore, regarding H and M as given quantities, there is one spheroidal form and only one, towards which the oscillating fluid mass continually approximates.

This discussion may be found in Laplace's *Mécanique Céleste*, tome ii. p. 61; Pontécoulant's *Système du Monde*, tome ii. p. 409; and in Tisserand's *Mécanique Céleste*, tome ii. p. 96.

125. Jacobi's Ellipsoid. It was discovered by Jacobi that an ellipsoid with three unequal axes is a possible form of relative equilibrium for a mass of rotating liquid.

If a mass of liquid revolves, as if rigid, about the axis of z with the angular velocity ω, and if X, Y, Z are the components of the attraction at the point (x, y, z), the equation of the free surface is

$$(X-\omega^2x)dx+(Y-\omega^2y)dy+Zdz=0.$$

Now, if the free surface is an ellipsoid,

$$X=Ax, \quad Y=By, \quad Z=Cz,$$

where A, B, C are independent of x, y, z.

Hence, if a, b, c are the semi-axes of the ellipsoid, we have if possible to identify the equations

$$(A-\omega^2)xdx+(B-\omega^2)ydy+Czdz=0,$$

$$\frac{x}{a^2}dx+\frac{y}{b^2}dy+\frac{z}{c^2}dz=0.$$

We must therefore satisfy the equations

$$A-\omega^2=\frac{\lambda}{a^2},\quad B-\omega^2=\frac{\lambda}{b^2},\quad C=\frac{\lambda}{c^2},$$

from which, by the elimination of λ and ω^2, we obtain

$$a^2b^2(B-A)-(a^2-b^2)c^2C=0 \quad . \quad . \quad . \quad (1)$$

Now, if $\quad D=\{(a^2+u)(b^2+u)(c^2+u)\}^{\frac{1}{2}},$

and if M is the mass of the liquid,

$$A=\tfrac{3}{2}M\int_0^\infty\frac{du}{(a^2+u)D},\qquad B=\tfrac{3}{2}M\int_0^\infty\frac{du}{(b^2+u)D},$$

$$C=\tfrac{3}{2}M\int_0^\infty\frac{du}{(c^2+u)D}.^*$$

The equation (1) then becomes

$$(a^2-b^2)\int_0^\infty\frac{du}{D}\left\{\frac{a^2b^2}{(a^2+u)(b^2+u)}-\frac{c^2}{c^2+u}\right\}=0.$$

If a is different from b, the relation between the axes must satisfy the equation

$$\int_0^\infty\frac{udu}{D^3}\left(\frac{1}{a^2}+\frac{1}{b^2}-\frac{1}{c^2}+\frac{u}{a^2b^2}\right)=0 \quad . \quad . \quad . \quad (2)$$

If a and b are given, this is an equation for determining c, and, since the left-hand member is negative when $c=0$, and positive when $c=\infty$, there must be one real value of c which satisfies the equation.

Since u/D^3 is positive, and since

$$\frac{1}{a^2}+\frac{1}{b^2}-\frac{1}{c^2}+\frac{u}{a^2b^2}$$

is positive if u is large enough, it follows that, when u is small, this last expression must be negative.

Hence it appears that

$$\frac{1}{c^2}>\frac{1}{a^2}+\frac{1}{b^2} \quad . \quad . \quad . \quad . \quad (3)$$

and therefore that c is less than the least of the two quantities a and b.

To find the angular velocity, we have

$$\omega^2(a^2-b^2)=Aa^2-Bb^2$$

$$=\tfrac{3}{2}M(a^2-b^2)\int_0^\infty\frac{udu}{(a^2+u)(b^2+u)D},$$

* See Kelvin and Tait's *Natural Philosophy*, Art. 494 *n*, or Minchin's *Statics*, vol. ii. p. 308.

and therefore, if a is different from b,

$$\omega^2 = \tfrac{3}{2}M\int_0^\infty \frac{udu}{(a^2+u)(b^2+u)D} \quad . \quad . \quad . \quad (4)$$

and, this expression being a positive quantity, a possible value of ω is obtained, and it is established that an ellipsoid with three unequal axes is a possible form of a mass of liquid rotating about the smallest axis.

126. That c must be the least axis may also be seen as follows:

$$\omega^2 = \frac{a^2A - c^2C}{a^2}$$

$$= \frac{3M}{2a^2}\int_0^\infty \left\{\frac{a^2}{a^2+u} - \frac{c^2}{c^2+u}\right\}\frac{du}{D}$$

$$= \frac{3M}{2a^2}(a^2-c^2)\int_0^\infty \frac{udu}{(a^2+u)(c^2+u)D},$$

which shows that for ω to be real, we must have $c<a$, and similarly $c<b$.

127. It was pointed out by Mr Todhunter, and demonstrated in the following manner, that the relative equilibrium of the rotating ellipsoid cannot subsist when the axis of rotation does not coincide with a principal axis.

Referred to the principal axes, let l, m, n be the direction cosines of the axis of rotation, M any point (x, y, z) of the mass, and N the foot of the perpendicular from M upon the axis.

Then $$ON = lx + my + nz,$$

and, if $ON = v$, the co-ordinates of N are lv, mv, nv.

The acceleration $\omega^2 MN$, when resolved parallel to the axes, gives rise to the components

$$\omega^2(x-lv), \quad \omega^2(y-mv), \quad \omega^2(z-nv);$$

therefore the differential equation of the free surface is

$$\{\omega^2(x-lv)-Ax\}dx + \{\omega^2(y-mv)-By\}dy + \{\omega^2(z-nv)-Cz\}dz = 0;$$

hence the form of the free surface is given by the equation

$$\omega^2(x^2+y^2+z^2) - \omega^2(lx+my+nz)^2 - Ax^2 - By^2 - Cz^2 = \text{constant},$$

and this cannot represent an ellipsoid referred to its principal axes, unless two of the quantities l, m, n vanish.

Mr Greenhill remarks that a particle of the liquid at the end of the axis of rotation will be at rest under the action of the attraction of the liquid alone, since the expression $\omega^2 r$ vanishes at that point.

Hence the attraction on the particle must be normal to the surface, which is only the case at the end of an axis.

128. We notice that if the mass of the fluid M be given, we have also an equation $\frac{4}{3}\pi\rho abc = M$, and this equation together with (2), (4) of Art. 125 may be regarded as determining a, b, c in terms of M, ρ, and ω.

These equations were investigated by C. O. Meyer,* and a full discussion will also be found in Tisserand's *Traité de Mécanique Céleste*, tome ii., chap. vii.,† showing that the maximum value of $\omega^2/2\pi\rho$ that will make a Jacobian ellipsoid a possible form of equilibrium is ·18709, and that for this particular value the ellipsoid is one of rotation coinciding with one of Maclaurin's spheroids. It is further shown that this value gives a unique maximum to the function on the right-hand side of equation (4) of Art. 125, and that for smaller values of $\omega^2/2\pi\rho$ there is one and only one ellipsoid.

To summarise our results relating to Maclaurin's spheroids and Jacobi's ellipsoids, we have:

if $\omega^2/2\pi\rho > \cdot 2247$, no spheroidal or ellipsoidal form,

if $\cdot 2247 > \omega^2/2\pi\rho > \cdot 18709$, two oblate spheroids,

and if $\cdot 18709 > \omega^2/2\pi\rho$, two oblate spheroids and one ellipsoid with three unequal axes.

129. It follows from Art. 125 (3) that the ellipticities of a Jacobian ellipsoid cannot be small, in fact that one of the axes is, in every case, at least $\sqrt{2}$ times the axis of rotation. In a complete discussion of the Jacobian ellipsoids containing numerical tables and diagrams,‡ Darwin remarks that the longer the ellipsoid the slower it rotates; that, while the angular velocity continually diminishes, the moment of momentum continually increases, and that the long ellipsoids are very nearly ellipsoids of revolution about an axis perpendicular to that of rotation.

130. Elliptic cylinder. It can also be shown that, theoretically, an elliptic cylinder is a possible form of the surface of an infinite mass of homogeneous gravitating liquid, rotating, as if rigid, about the axis of the cylinder.

* *Crelle's Journal*, tome xxiv. (1842).

† For an abstract of the analysis see Appell, *Traité de Mécanique Rationnelle*, tome iii. p. 170.

‡ "On Jacobi's Figure of Equilibrium for a rotating mass of fluid," *Proc. Royal Soc.*, vol. xli. (1887), p. 319; or *Scientific Papers*, vol. iii. p. 119.

If a and b are the semiaxes, the components of the attraction at the internal point x, y are

$$\frac{4\pi\rho bx}{a+b} \text{ and } \frac{4\pi\rho ay}{a+b}$$

(Kelvin and Tait, Art. 494 p), and the equation of the free surface is therefore

$$\left(\frac{4\pi\rho b}{a+b}-\omega^2\right)xdx+\left(\frac{4\pi\rho a}{a+b}-\omega^2\right)ydy=0.$$

Identifying this equation with

$$\frac{xdx}{a^2}+\frac{ydy}{b^2}=0,$$

we find that

$$\omega^2=4\pi\rho ab/(a+b)^2.$$

This determines ω when ρ, a, b are given; but if ω, ρ are given we see that since

$$\frac{a-b}{a+b}=\sqrt{1-\frac{\omega^2}{\pi\rho}}$$

an elliptic cylinder will not be a possible form of equilibrium unless $\omega^2<\pi\rho$.

131. Poincaré's Theorem. We have seen that a Jacobian ellipsoid is an impossible form of relative equilibrium if

$$\omega^2/2\pi\rho>{\cdot}18709,$$

an oblate spheroid is impossible if $\omega^2/2\pi\rho>{\cdot}2247$, and an elliptic cylinder is not a possible form if $\omega^2/2\pi\rho>{\cdot}5$; Poincaré has proved that *if* $\omega^2/2\pi\rho>1$ *there is no figure of equilibrium possible.** For a necessary condition of equilibrium is that at every point of the free surface the resultant of the attraction and centrifugal force should be directed towards the interior, otherwise a part would be detached. Let V be the potential of the attracting forces and r the distance from the axis, and let

$$U=V+\tfrac{1}{2}\omega^2r^2.$$

The resultant outward normal force is $\frac{\partial U}{\partial n}$ and, for equilibrium, at every point of the free surface $\frac{\partial U}{\partial n}$ must be negative. By Green's Theorem $\iint\frac{\partial U}{\partial n}dS=\iiint\nabla^2U dxdydz$, where the first integral is taken

* *Bulletin Astron.*, tome ii. p. 117, or *Figures d'équilibre d'une masse fluide*, p. 11.

over the surface and the second throughout the volume of the fluid. And

$$\nabla^2 U = \nabla^2 V + 2\omega^2 = -4\pi\rho + 2\omega^2.$$

Therefore $$\iint \frac{\partial U}{\partial n} dS = 2(\omega^2 - 2\pi\rho) \times \text{volume},$$

and if $\omega^2 > 2\pi\rho$, the left-hand member is positive, which implies that at some points on the surface the resultant force is directed outwards and therefore equilibrium is impossible.

132. Other equilibrium forms. In addition to the forms that we have considered, the annulus was first considered by Laplace * in connection with the theory of Saturn's rings, and has since been the subject of much investigation.

In the second edition of Kelvin and Tait's *Natural Philosophy*, § 778″, a number of results relating to the stability of the forms already discussed were announced without proof. In attempting to establish these results, Poincaré was led to write a celebrated paper which appeared in the *Acta Mathematica*, 7, Stockholm, 1885. In this paper the problem of figures of equilibrium is discussed in a more general manner. It is shown that possible figures of equilibrium form linear series, that is, series depending on a single parameter, such as the angular velocity, and such that to each value of the parameter corresponds either one and one only, or else a finite number of figures, and such that these figures vary in a continuous manner when the parameter is varied. Thus the Maclaurin's spheroids form a linear series, and Jacobi's ellipsoids form another. It may happen that the same figure belongs to two distinct linear series ; such a figure is called a form of "bifurcation." Thus there is a particular member of the series of spheroids which at the same time belongs to the series of Jacobi's ellipsoids. Poincaré also considered, in this paper, the question of the stability of forms of equilibrium, and showed that if a series of figures are stable up to a form of bifurcation, then beyond that point the figures are unstable, the stable figures now belonging to the other series involved in the form of bifurcation. Thus Maclaurin's spheroid is stable only so long as its eccentricity is less than ·8127, which is the point of bifurcation, and at this point Jacobi's ellipsoids become stable. In attempting to find points of bifurcation in the series of Jacobi's

* *Mécanique Céleste*, tome ii. p. 155. See also Tisserand, *Mécanique Céleste*, tome ii. chaps. ix. x. xii., where the researches of Laplace, Clerk Maxwell, and Mme. Kowalewski are discussed.

ellipsoids by the use of Lamé's functions, Poincaré found that there are an infinite number of series of figures of equilibrium. All the figures are symmetrical with regard to a plane perpendicular to the axis of rotation; they all have at least one plane of symmetry passing through the axis and some of them are figures of revolution. Among these figures only one is stable and it has only two planes of symmetry; it is the form that arises from the first bifurcation in the series of Jacobi's ellipsoids and has been called the pear-shaped figure of equilibrium, because of the resemblance to a pear of the figure sketched in Poincaré's paper.* Further investigation, however, has shown that the true form has less resemblance to a pear than was at first supposed; it has been discussed by Darwin in two papers,† and its form determined to a second approximation. At the point of bifurcation the axes of the Jacobian ellipsoid are as 65066 : 81498 : 188583, and $\omega^2/2\pi\rho = \cdot 14200$; and the pear-shaped figure represents a small departure from this Jacobian ellipsoid,

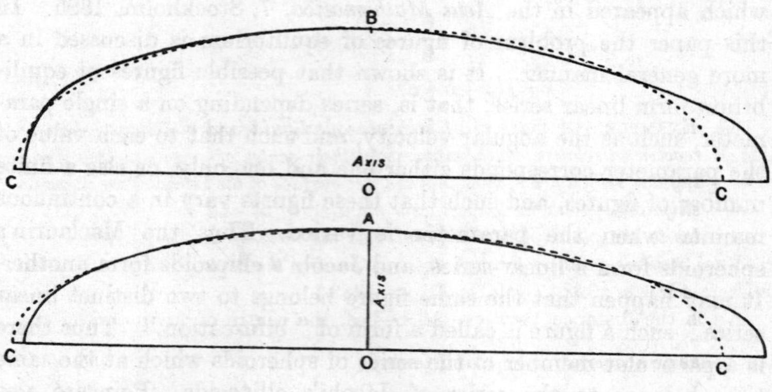

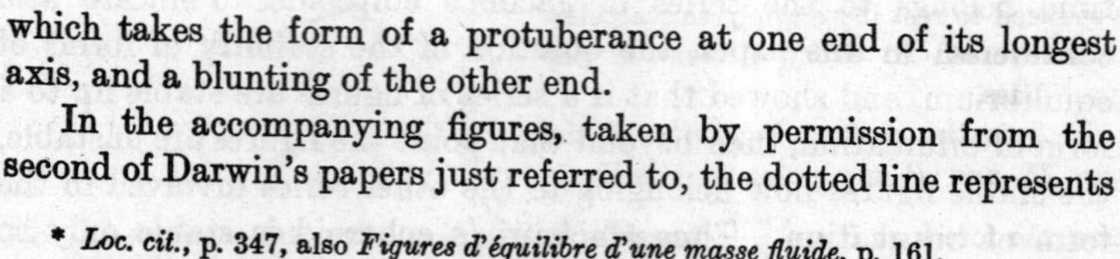

which takes the form of a protuberance at one end of its longest axis, and a blunting of the other end.

In the accompanying figures, taken by permission from the second of Darwin's papers just referred to, the dotted line represents

* *Loc. cit.*, p. 347, also *Figures d'équilibre d'une masse fluide*, p. 161.

† "On the pear-shaped figure of equilibrium of a rotating mass of liquid," *Phil. Trans.*, vol. 198 A (1901), p. 301, or *Scientific Papers*, vol. iii. p. 288; and "The stability of the pear-shaped figure of equilibrium of a rotating mass of liquid," *Phil. Trans.*, vol. 200 A (1902), p. 251, or *Scientific Papers*, vol. iii. p. 317. For a simple account of the stability of these figures see also an interesting paper by the same author on "The Genesis of Double Stars," being chap. xxviii. in the volume *Darwin and Modern Science*.

the Jacobian ellipsoid, and the other curve the pear-shaped figure; the upper is the equatorial section, and the lower is the meridional section in the plane of symmetry.

133. The following expressions for the attraction of a solid homogeneous ellipsoid of small ellipticities are often of use in discussing the forms assumed by masses of rotating liquid; viz. if a, b, c, the semiaxes, are such that $b=a(1-\epsilon)$ and $c=a(1-\eta)$, then the component attractions at an internal point (x, y, z) are

$$A\rho x, \quad B\rho y, \quad C\rho z,$$

where

$$A=\tfrac{4}{3}\pi(1-\tfrac{2}{5}\epsilon-\tfrac{2}{5}\eta),$$
$$B=\tfrac{4}{3}\pi(1+{}^{4}\epsilon-\tfrac{2}{5}\eta),$$
$$C=\tfrac{4}{3}\pi(1-\tfrac{2}{5}\epsilon+\tfrac{4}{5}\eta).*$$

These expressions may also be written in the symmetrical form

$$A=\tfrac{4}{3}\pi\left(1-\frac{2}{5}\,\frac{2a-b-c}{a}\right), \text{ etc.}$$

or as

$$A=\tfrac{4}{3}\pi\left(1-\frac{6}{5}\,\frac{a-k}{k}\right), \text{ etc.}$$

where

$$k=\tfrac{1}{3}(a+b+c).$$

134. EXAMPLE. *A mass m of homogeneous liquid and a distant sphere of mass M revolve in relative equilibrium about their centre of gravity with a small uniform angular velocity ω; show that the free surface of the liquid is an ellipsoid of small ellipticities with its longest axis pointing to M and its smallest axis at right angles to the plane of motion, and that the ratio of the ellipticities of the principal sections passing through the line joining the centres of gravity of the bodies is $4M+m : 3M$.*† (Math. Tripos, 1888.)

If d is the distance between the bodies, the centre of gravity O of the mass m has an acceleration $\dfrac{\mu M}{d^2}$, and O may be reduced to rest if we apply this acceleration reversed to every element of the liquid mass.

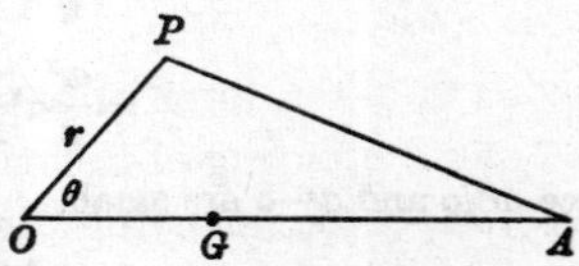

If A is the centre of gravity of the mass M, and P any point in the liquid mass, the forces at P are $\dfrac{\mu M}{PA^2}$ towards A, $\dfrac{\mu M}{AO^2}$ parallel to AO, the force due to the self-attraction of the liquid and the centrifugal force. Now $\dfrac{\mu M}{PA^2}$ along PA is equivalent to $\dfrac{\mu M}{PA^3}\cdot PO$ along PO and $\dfrac{\mu M}{PA^3}\cdot OA$ parallel to OA.

---

* See Routh's *Analytical Statics*, vol. ii. § 221 (2nd edition).

† Problems of this class were discussed by Laplace in the third book of the *Mécanique Céleste.*

The former

$$=\frac{\mu Mr}{\{d^2+r^2-2dr\cos\theta\}^{\frac{3}{2}}}=\frac{\mu Mr}{d^3}$$

to the first order of r/d.

The latter combined with $\frac{\mu M}{AO^2}$

$$=\frac{\mu Md}{\{d^2+r^2-2dr\cos\theta\}^{\frac{3}{2}}}-\frac{\mu M}{d^2}=\frac{\mu M}{d^2}\left\{1+\frac{3r}{d}\cos\theta-1\right\}$$

$$=\frac{3\mu Mr\cos\theta}{d^3}$$

parallel to OA.

If we assume an ellipsoidal form and take the axis of x along OA, and Oz for axis of rotation, we have

$$\frac{dp}{\varrho}=\omega^2(xdx+ydy)-A\varrho xdx-B\varrho ydy-C\varrho zdz-\frac{\mu Mr}{d^3}dr+\frac{3\mu Mx}{d^3}dx.$$

And the free surface must be of the form

$$x^2\left(\omega^2-A\varrho+\frac{3\mu M}{d^3}-\frac{\mu M}{d^3}\right)+y^2\left(\omega^2-B\varrho-\frac{\mu M}{d^3}\right)-z^2\left(C\varrho+\frac{\mu M}{d^3}\right)=\text{const.}$$

$$\therefore a^2\left(A\varrho-\frac{2\mu M}{d^3}-\omega^2\right)=b^2\left(B\varrho+\frac{\mu M}{d^3}-\omega^2\right)=c^2\left(C\varrho+\frac{\mu M}{d^3}\right).$$

Now since the masses are rotating about their centre of gravity G with angular velocity ω,

$$\therefore \omega^2\cdot OG=\frac{\mu\cdot M}{d^2},$$

but $$(M+m)OG=Md;$$

$$\therefore \omega^2=\frac{\mu(M+m)}{d^3};$$

$$\therefore a^2A-b^2B=\frac{\omega^2}{\varrho}\left\{a^2\left(1+\frac{2M}{M+m}\right)-b^2\left(1-\frac{M}{M+m}\right)\right\}$$

$$=\frac{\omega^2}{\varrho}a^2\frac{3M}{M+m}.$$

since ω^2/ϱ and $a-b$ are small.

So also $$a^2A-c^2C=\frac{\omega^2}{\varrho}\left\{a^2\left(1+\frac{2M}{M+m}\right)+c^2\frac{M}{M+m}\right\}$$

$$=\frac{\omega^2}{\varrho}a^2\frac{4M+m}{M+m}.$$

But from the last Article,

$$a^2A-b^2B=\tfrac{4}{3}\pi\left\{(a^2-b^2)-\tfrac{6}{5}a^2\frac{(a-k)}{k}+\tfrac{6}{5}b^2\frac{(b-k)}{k}\right\}$$

$$=\tfrac{4}{3}\pi(a-b)\left\{a+b-\tfrac{6}{5}\left(\frac{a^2+ab+b^2}{k}-a-b\right)\right\},$$

and to get a result correct to the first order of the small difference $a-b$ we may put $k=b=a$ in the last factor, so that

$$a^2A-b^2B=\tfrac{16}{15}\pi a(a-b).$$

Similarly
$$a^2A-c^2C=\tfrac{16}{15}\pi a(a-c).$$

Hence
$$\frac{a-b}{a-c}=\frac{a^2A-b^2B}{a^2A-c^2C}=\frac{3M}{4M+m}.$$

EXAMPLES

1. A thin spherical shell of radius a is just not filled with gravitating liquid of density ϱ. If the liquid be rotating in relative equilibrium with angular velocity ω about a diameter, prove that the tension in the shell across the great circle at right angles to the axis of rotation is at any point in that circle equal to $\omega^2\varrho a^3/8$.

2. A mass of liquid of density ϱ_1 is surrounded by a mass of liquid of density ϱ and the whole completely fills a case in the form of an oblate spheroid of small ellipticity ε; if the case rotates about its axis with small uniform angular velocity ω, prove that a possible form of the common surface is an oblate spheroid of ellipticity ε_1 given by

$$15\omega^2/16\pi=\varepsilon_1\varrho_1+\tfrac{3}{2}(\varepsilon_1-\varepsilon)\varrho.$$

3. A case in the form of a *prolate* spheroid of small ellipticity ε is filled by a fluid nucleus of density $\varrho+\sigma$ surrounded by a fluid of density ϱ. Show that, if it rotates round its axis of figure with angular velocity $\left(\frac{8}{5}\pi\varrho\varepsilon\right)^{\frac{1}{2}}$, a possible form of the common surface is a sphere.

4. A mass of homogeneous liquid of density ϱ completely fills a case in the form of the ellipsoid $x^2/a^2+y^2/b^2+z^2/c^2=1$, and rotates as a rigid body about the line $x/l=y/m=z/n$ with uniform angular velocity ω; show that if $\frac{1}{2}\lambda\varrho$ is the greatest excess of the pressure at the centre over the pressure at a point on the surface,

$$\frac{l^2}{\dfrac{1}{A-\lambda/a^2}-\dfrac{1}{\omega^2}}+\frac{m^2}{\dfrac{1}{B-\lambda/b^2}-\dfrac{1}{\omega^2}}+\frac{n^2}{\dfrac{1}{C-\lambda/c^2}-\dfrac{1}{\omega^2}}=0,$$

where Ax, By, Cz are the components of the attraction at an internal point.

5. Two gravitating liquids which do not mix, and whose densities are ϱ, $\sigma(\varrho>\sigma)$, are enclosed in a rigid spherical envelope, and the whole rotates in relative equilibrium with a small uniform angular velocity ω about a diameter of the sphere. Show that a possible form of the common surface of the two liquids is an oblate spheroid of ellipticity $\frac{15}{16}\,\omega^2/\pi(\varrho+\frac{3}{2}\sigma)$.

6. A given mass of gravitating fluid of density ϱ can rotate in relative equilibrium with angular velocity Ω with its free surface in the form of an ellipsoid with three unequal axes, the greatest semiaxis being a. A rigid vessel of this form is now made and the fluid in it is set rotating with the vessel in relative equilibrium with angular velocity ω about the least axis. Prove that the pressure at any point of the surface is

$$\tfrac{1}{2}\varrho(\omega^2-\Omega^2)(x^2+y^2) \text{ or } \tfrac{1}{2}\varrho(\omega^2-\Omega^2)(x^2+y^2-a^2),$$

according as ω is greater or less than Ω.

7. A solid sphere of mean density ϱ is covered by a thin layer of liquid of uniform density σ. The whole rotates with small uniform angular velocity ω about an axis through the centre of the sphere; the solid sphere attracts according to the law of the inverse square as if concentrated at a point on the axis at a small distance c from its centre, and the liquid also attracts according to the law of the inverse square. Show that the outer surface of the liquid is approximately a spheroid of ellipticity $15\omega^2/8\pi(5\varrho-3\sigma)$, with its centre at a distance $\varrho c/(\varrho-\sigma)$ from the centre of the sphere.

8. A solid gravitating sphere of radius a and density ϱ is surrounded by a gravitating liquid of volume $\frac{4}{3}\pi(b^3-a^3)$ and density σ. The whole is made to rotate with small angular velocity ω. Show that the form of the free surface of the liquid is the spheroid of small ellipticity ε given by

$$r=b(1-\tfrac{2}{3}\varepsilon P_2),$$

where
$$\varepsilon=\frac{15\omega^2 b^3}{8\pi\{5(\varrho-\sigma)a^3+2\sigma b^3\}},$$

and P_2 is Legendre's coefficient of the second order.

9. A homogeneous gravitating fluid just does not fill a rigid envelope in the form of an oblate ellipsoid. The fluid is rotating in relative equilibrium round the polar axis with kinetic energy E. If it rotates with kinetic energy E_1 the envelope is a free surface of zero pressure. Prove that, for all values of E whether greater or less than E_1, the tension per unit length across the equatorial section of the envelope is

$$\frac{15}{32}\frac{E\sim E_1}{A},$$

where A is the area of a polar section of the ellipsoid.

10. A nearly spherical solid of mass M, the equation to whose surface is $r=a(1+aP_4)$, has a mass m of liquid on its surface, the solid and liquid attracting according to the Newtonian law, and the whole rotates about the axis of the harmonic with angular velocity ω. Show that the equator will be uncovered if $m<9aM/(12\lambda-4)-5\omega^2a^3/(10\lambda-6)$, and that the poles will be uncovered if $m<6aM/(3\lambda-1)+5\omega^2a^3/(5\lambda-3)$, where λ is the ratio of the density of the solid to that of the liquid.

11. Assuming the Earth to consist of a fluid surrounding a solid spherical nucleus, prove that the ellipticity, supposed small, is given by

$$\varepsilon=m\frac{D/\varrho}{4/5+2(D/\varrho-1)},$$

where m is the ratio of the centrifugal force at the equator to the gravity there, D is the mean density of the whole Earth, and ϱ the density of the fluid.